T0224916

Ethics for Bioengineers

Synthesis Lectures on Biomedical Engineering

Editor
John D. Enderle, *University of Connecticut*

Lectures in Biomedical Engineering will be comprised of 75- to 150-page publications on advanced and state-of-the-art topics that spans the field of biomedical engineering, from the atom and molecule to large diagnostic equipment. Each lecture covers, for that topic, the fundamental principles in a unified manner, develops underlying concepts needed for sequential material, and progresses to more advanced topics. Computer software and multimedia, when appropriate and available, is included for simulation, computation, visualization and design. The authors selected to write the lectures are leading experts on the subject who have extensive background in theory, application and design.

The series is designed to meet the demands of the 21st century technology and the rapid advancements in the all-encompassing field of biomedical engineering that includes biochemical, biomaterials, biomechanics, bioinstrumentation, physiological modeling, biosignal processing, bioinformatics, biocomplexity, medical and molecular imaging, rehabilitation engineering, biomimetic nano-electrokinetics, biosensors, biotechnology, clinical engineering, biomedical devices, drug discovery and delivery systems, tissue engineering, proteomics, functional genomics, molecular and cellular engineering.

Ethics for Bioengineers
Monique Frize
2011

Computational Genomic Signatures
Ozkan Ufuk Nalbantoglu and Khalid Sayood
2011

Digital Image Processing for Ophthalmology: Detection of the Optic Nerve Head
Xiaolu Zhu, Rangaraj M. Rangayyan, and Anna L. Ells
2011

Modeling and Analysis of Shape with Applications in Computer-Aided Diagnosis of Breast Cancer
Denise Guliato and Rangaraj M. Rangayyan
2011

Basic Feedback Controls in Biomedicine
Charles S. Lessard
2009

Understanding Atrial Fibrillation: The Signal Processing Contribution, Part I
Luca Mainardi, Leif Sörnmo, and Sergio Cerutti
2008

Understanding Atrial Fibrillation: The Signal Processing Contribution, Part II
Luca Mainardi, Leif Sörnmo, and Sergio Cerutti
2008

Introductory Medical Imaging
A. A. Bharath
2008

Lung Sounds: An Advanced Signal Processing Perspective
Leontios J. Hadjileontiadis
2008

An Outline of Informational Genetics
Gérard Battail
2008

Neural Interfacing: Forging the Human-Machine Connection
Susanne D. Coates
2008

Quantitative Neurophysiology
Joseph V. Tranquillo
2008

Tremor: From Pathogenesis to Treatment
Giuliana Grimaldi and Mario Manto
2008

Introduction to Continuum Biomechanics
Kyriacos A. Athanasiou and Roman M. Natoli
2008

The Effects of Hypergravity and Microgravity on Biomedical Experiments
Thais Russomano, Gustavo Dalmarco, and Felipe Prehn Falcão
2008

Ethics for Bioengineers

Monique Frize

ISBN: 978-3-031-00518-3   paperback
ISBN: 978-3-031-01646-2   ebook

DOI 10.1007/978-3-031-01646-2

A Publication in the Springer series
SYNTHESIS LECTURES ON BIOMEDICAL ENGINEERING

Lecture #42
Series Editor: John D. Enderle, *University of Connecticut*
Series ISSN
Synthesis Lectures on Biomedical Engineering
Print 1930-0328   Electronic 1930-0336

Ethics for Bioengineers

Monique Frize

Carleton University
University of Ottawa

SYNTHESIS LECTURES ON BIOMEDICAL ENGINEERING #42

ABSTRACT

Increasingly, biomedical scientists and engineers are involved in projects, design, or research and development that involve humans or animals. The book presents general concepts on professionalism and the regulation of the profession of engineering, including a discussion on what is ethics and moral conduct, ethical theories and the codes of ethics that are most relevant for engineers. An ethical decision-making process is suggested. Other issues such as conflicts of interest, plagiarism, intellectual property, confidentiality, privacy, fraud, and corruption are presented. General guidelines, the process for obtaining ethics approval from Ethics Review Boards, and the importance of obtaining informed consent from volunteers recruited for studies are presented. A discussion on research with animals is included.

Ethical dilemmas focus on reproductive technologies, stem cells, cloning, genetic testing, and designer babies. The book includes a discussion on ethics and the technologies of body enhancement and of regeneration. The importance of assessing the impact of technology on people, society, and on our planet is stressed. Particular attention is given to nanotechnologies, the environment, and issues that pertain to developing countries. Ideas on gender, culture, and ethics focus on how research and access to medical services have, at times, been discriminatory towards women. The cultural aspects focus on organ transplantation in Japan, and a case study of an Aboriginal child in Canada; both examples show the impact that culture can have on how care is provided or accepted. The final section of the book discusses data collection and analysis and offers a guideline for honest reporting of results, avoiding fraud, or unethical approaches. The appendix presents a few case studies where fraud and/or unethical research have occurred.

KEYWORDS

ethical theories, codes ethics, ethics approval, consent, impact on society, ethical dilemmas, good research practices, data collection and analysis

Contents

Preface

Before becoming a professor of electrical, systems, computer, and biomedical engineering, I was a clinical engineer, managing medical technologies and patient safety in a large hospital in Montreal, Canada between 1971 and 1979, then for seven hospitals in New-Brunswick. In this role, I saw clearly the need for biomedical engineers (and engineers in general) to be knowledgeable about ethical guidelines, codes of ethics, and understanding the impact of technology on health care and on society. So when I became a professor in December 1989, my first choice of course to teach was Ethics and Professional Practice at the University of New-Brunswick. I re-designed the course so that students develop an understanding of a number of ethical theories and codes of ethics, and learn how to make decisions when facing an ethical issue or dilemma. They also learned how to think of the impact of technology on society when they would be involved in design or developing technology.

Later, at Carleton University and the University of Ottawa, I adapted the course for graduate students in biomedical engineering. Thus I added discussions on ethical dilemmas that are pertinent for this field of study, and students learned how to apply for ethical clearance to an Ethics Review Board for projects that involve humans or animals. Some of the discussion pertained to issues that are prevalent in developing countries. The course ended with a review of good research practices, with a focus on data collection and analysis. After a decade of teaching this material, I thought it would be useful to put these ideas in a book for instructors who wish to teach these concepts in their institution.

In the past few years, books on ethics have been published, but for my course, several had to be consulted to cover the material I wanted to convey to students in this field. This book consolidates material to support the course. It can be also useful for anyone involved in design, development, or research, whether in industry, hospitals, government, or in universities and colleges.

I am grateful to all the students who have provided interesting discussions in their essays and who participated actively in the development of the ideas that have contributed to improving the course over the years. My hope is that they will be responsible engineers who design and develop technological solutions with people, society, and our world in mind.

Monique Frize
November 2011

Introduction

Increasingly, biomedical scientists and engineers work on projects, design, or research and development that involve humans or animals. Therefore, it is critical that students, engineers, and scientists acquire an in-depth knowledge of ethical considerations prior to developing a plan for the studies to be performed. The book covers the following material: The first chapter begins with general concepts on professionalism and the regulation of the profession of engineering; it includes a discussion on what is ethics and moral conduct, presents ethical theories and codes of ethics that are most relevant for engineers, and suggests an ethical decision-making process that should be followed. Other issues such as conflicts of interest, plagiarism, intellectual property, confidentiality, privacy, fraud, and corruption are presented.

In Chapter 2, the discussion focuses on the procedure to follow when planning to do some testing on human subjects or animals. This includes becoming aware of guidelines that must be followed and using the process to apply for ethical approval to the Ethics Review Board of the institution where the work is to be carried-out. This includes becoming familiar with application forms, learning what is expected to be presented in the proposal, and obtaining informed consent from volunteers who will be recruited. All projects must consider the principles of autonomy, beneficence, non-maleficence, and justice. The chapter concludes with a discussion on work or research with animals.

Chapter 3 presents examples of ethical dilemmas in biomedical research and the importance of thinking of all sides of the issues. The main example presented is on reproductive technologies and concludes with other issues such as stem cells, cloning, genetic testing, and designer babies. This list is not exhaustive and several web sites can be visited to find literally hundreds of topics.

Chapter 4 discusses the importance of assessing the impact of technology on people, society, and our planet. In all developments of science and technology, there are positive and negative impacts which need to be identified if the projects and developments are to minimize the negative impacts. Particular attention is given to nanotechnologies, as their development has been far more rapid than studies on their safety, social or ethical impacts. Issues that are prevalent in developing countries are mentioned.

Chapter 5 presents ideas on gender, culture, and ethics. In the first part, the focus is on how research and access to medical services have, in the past, been discriminatory towards women. This pertains to the principle of justice, and although much progress has been made by granting agencies to insure inclusion, there are still areas where discrimination still exists. The cultural aspects focus on organ transplantation in Japan, and a case study of an Aboriginal child in Canada. Both examples show how culture can have an impact on how care is provided or accepted.

2 INTRODUCTION

Chapter 6 discusses data collection and analysis. This includes general principles for honest reporting of results, avoiding fraud and unethical approaches.

CHAPTER 1

Introduction to Ethics

1.1 THE ENGINEERING PROFESSION AND ITS REGULATION

The Oxford Dictionary defines profession as a paid occupation, especially one that involves prolonged training and formal qualifications. Main characteristics of a profession are the following:

> First, a profession is more than an occupation; it is a career or vocation… Second, because professionals provide valuable goods and services, they have public responsibilities and can be held publicly accountable. Third, although professionals have public responsibilities, they are also granted a great deal of autonomy; society allows professionals to be self-regulating…Fourth, ethical standards play a key role in professional conduct by promoting self-regulation and public responsibility [Resnick and Shamoo, 2003].

The statement above is true for some countries but not for all. Self-regulation implies that the professionals develop their standards and rules and have a mechanism for punishing members who are found guilty of misconduct, or incompetence, or are in breach of the Code of Ethics of their profession. In Canada and in the United States of America, the engineering profession is self-regulated; it is governed by its own association of engineers, and therefore by the members. In order to join the association in a province or territory in Canada, applicants have to meet the following criteria: have graduated from an accredited engineering program or from one that is recognized as equivalent, or else they will be screened for competence through a series of exams. In addition, they must be of age (18 or 19, depending on the jurisdiction); they must be a citizen or have permanent resident status in Canada; be of good character, as confirmed by references; successfully complete the professional practice exam; and have the required number of years of relevant engineering work experience, also verified by referees. The professional practice exam is usually composed of two parts: on ethical principles and code of ethics; and on principles of contract law. Applicants are encouraged to join as an Engineer-in-Training (EIT) and obtain their Professional Engineer designation (P.Eng.) after they have met all the criteria listed above. Although there is a national association of engineers in the two countries listed above, they are an umbrella organization. The role of Engineers Canada is as follows:

> It accredits Canadian undergraduate engineering programs that meet the profession's high education standards. Graduates of those programs are deemed by the profession to have the required academic qualifications to be licensed as professional engineers

in Canada. It also assesses the equivalency of the accreditation systems used in other nations relative to the Canadian system, and monitors the accreditation systems employed by the engineering bodies which have entered into mutual recognition agreements with Engineers Canada. Engineers Canada develops national guidelines on the qualifications, standards of practice, and ethics expected of professional engineers. It also publishes the Engineers Canada Examination Syllabus and the Engineers Canada List of Foreign Engineering Educational Institutions and Professional Qualifications [Engineers Canada, 2008].

In the United States, the licensure process can vary slightly from state to state but in general, the steps are as follows: The new graduate applies to become an engineer-in-training and successfully completes a "Fundamentals of Engineering" exam. Four years of experience at the appropriate level is required to be licensed as a professional engineer (P.E.); candidates are expected to learn the specific licensure requirement of the state or territory where they intend to practice, and write the "Principles and Practice of Engineering" exam of the jurisdiction they have chosen for their practice. In both countries, the national organization provides resources to help applicants succeed in their licensure process [National Society of Professional Engineers, 2011].

An example of multinational agreements is the Washington Accord, signed in 1989 by twelve countries including Canada and the United States, which recognizes substantial equivalence in the accreditation of qualifications in professional engineering, normally of four years duration. This was to enhance mobility for graduates of engineering programs which met the quality standards established by the signing nations. Mobility is desirable, and there are some major efforts made by the various associations to minimize the obstacles to this principle [International Engineering Alliance, 2011]. The designation of professional engineer (P.E. in the USA, ing. in the Province of Quebec, and P. Eng. in the rest of Canada) is a responsibility of each state, province, or territory. In other regions of the world, the profession is regulated by the government, as in France and Sweden. In North America, each professional engineering association and technical society provides its own ethical code and guideline to help engineers avoid misconduct, negligence, incompetence, or corruption. A complaint against an engineer can lead to discipline, which can include a fine, and/or losing the license to practice. Knowledge of the ethical decision-making process can guide engineers facing complex and difficult moral dilemmas. Principles of ethics and ethical codes are at the core of the duties and responsibilities of engineers. Although there are small differences between the ethical codes of various jurisdictions, many of the core values are the same. Engineers should review regularly the code of ethics of the jurisdiction in which they practice to ensure that they remain ethical in all their professional activities and keep up with changes in the codes when this occurs. Ethics is a dynamic concept which must be adapted to emerging issues as they arise.

1.2   A GUIDE TO MORAL CONDUCT

Engineers have access to several guides to ensure that they act ethically in all their professional activities. This helps prevent engineers being in a situation of misconduct, incompetence, fraud, corruption, or of committing a crime. First and foremost are the laws in the jurisdiction where the engineer practices. But this is not enough to ensure ethical behavior. Kluge argues that laws are prescriptive; they tell people what to do and not to do. Laws regulate the behavior of the members of that society; they are decided by the people designated to that role and so to some extent are arbitrary [Kluge, E., 2005]. Laws have to be obeyed of course, but they alone cannot guide ethical behavior; laws have not always been ethical; take for example the law that allowed slavery, or that women were non-persons, or the nonconsensual sterilization of severely mentally handicapped persons [Kluge, E., 2005]. Moreover, several aspects of ethical behavior are not inscribed in laws. One has to be aware of ethical theories and ethical codes that help guide behavior and actions. The third source of guidance for actions, especially as they pertain to designing, testing, or research that involve humans are the four principles of autonomy, beneficence, non-maleficence, and justice. More will be said about these concepts later.

The definition of morality can be found in dictionaries, Wikipedia, and many education web sites. Gert, B. [2011] in the *Stanford Encyclopedia of Philosophy*, argues that "[t]he term 'morality' can be used either descriptively, to refer to some codes of conduct put forward by a society, or some other group, such as a religion, or accepted by an individual for his/her own behavior; or normatively to refer to a code of conduct that, given specified conditions, would be put forward by all rational persons." Morality is what people believe to be right and good and the reasons for it. Although the label of right and wrong can vary according to circumstances, there are some universal rules that must guide humans at all times, such as not killing, stealing, or lying. But in some cases, there is no universally correct way to behave and the choice of how to act must follow an analysis of alternatives and consequences derived from each choice. This is where ethical theories can help in assessing what would be the most ethical behavior when facing a moral dilemma.

Ethics is the philosophical study of morality, a rational examination into moral beliefs and behaviors, the study of right and wrong, of good and evil in human conduct. Since bioengineers and scientists will undoubtedly be involved in projects that involve humans or animals, it is essential to be informed on all aspects of ethical conduct prior to planning such projects. It is also important to be aware of the specific guidelines of the institution where the work is to be carried-out and be familiar with the application process for obtaining a certificate that will allow the study to proceed. For clinical engineers working in health care institutions, there exist a number of articles discussing ethics specifically aimed at this group of bioengineers [Goodman, G., 1989, Saha and Saha, 1986]. The following section presents a number of ethical theories and codes of ethics most relevant for bioengineers.

1.3   ETHICAL THEORIES

There exist many ethical theories, some of them dating from the time of Socrates, Plato and Aristotle. Ethical theories have been classified by philosophical textbooks into three categories: Meta-Ethics, Normative Ethics, and Applied Ethics. The ethical theories presented here are in the Normative Ethics category. In this section, we focus on the theories which have more relevance to ethical decision-making in engineering. A theory such as Subjective Relativism has limited utility in ethical decision-making, as it encourages decisions based on an individual's perspective. As its name suggests, persons can have completely opposite views, but both can be considered to be right [Quinn, M., 2005]. The same thought applies to Cultural Relativism where the meaning of right and wrong rests with a society's moral guidelines. Once again, much latitude is provided to this subjective approach and this is not very useful for decision-making when facing a moral dilemma in engineering. The Divine Command Theory is based on particular religious beliefs, which can vary from one religion to the other, and is based on old manuscripts which have not addressed modern issues created by the internet and computers. These theories are not based on rational thinking and thus have a limited value for our purpose and will not be discussed in more detail.

On the other hand, theories such as Kantianism, Act Utilitarianism, Rule Utilitarianism, Social Contract, Rights Theory, and Rawls' Theory of Justice are based on rational reasoning and appear to be most helpful for ethical decision-making for bioengineers [Quinn, M., 2005].

Kantianism : Immanuel Kant (1724–1804), a philosopher of the 18th century, based his ethical theory on the principle of duty. The theory is referred to as the *categorical imperative*, since it pertains to actions that are universal and which arise from the sense of duty. We are compelled to act in certain ways because of some moral rules. In fact, several of these rules are also written in laws such as be honest, be fair, do not hurt others, keep promises, and obey the law. These rules in turn lead to respect for others. In his first formulation, Kant defines the categorical imperative: *"Act only from moral rules that you can, at the same time, will to be universal moral laws"* [Quinn, M., 2005].

Kant's second formulation states: *"Act so that you always treat both yourself and other people as ends in themselves and never only as a means to an end"* [Quinn, M., 2005]. In other words, it is not right to use others for our own purposes. There are examples of researchers who did not tell the whole truth to human subjects about the experiments they were going to perform on them and thus used their subjects to reach their own ends. This is a breach of ethics according to Kant's theory. All interactions must be respectful of the humanity of others. Kantianism is rational because logic is used to explain why a solution to an ethical problem has been chosen. This theory produces universal moral guidelines that apply to all people for all times. All persons must be treated as moral equals; that is, people in similar situations should be treated equally. Kantianism provides a framework to combat discrimination. A weakness of this theory is that it does not apply well to cases that are ambivalent or when there is a conflict between rules. Kantianism does not allow any exceptions to moral laws. This theory mainly supports moral decision-making based on logical reasoning from facts and commonly held views. It is a workable theory for ethical decision-making in cases where universal principles are concerned [Andrews, G., 2005, Quinn, M., 2005].

Act Utilitarianism (Act U): fits in the category of consequential theories, which means that one must analyze the outcomes or consequences of an act. It was first formulated by Jeremy Bentham (1748–1832) and later developed by John Stuart Mill (1806–1873) into Rule Utilitarianism (discussed below). Act Utilitarianism is based on the principle of utility. It states that an action is good if it benefits someone and bad if it harms someone. This is also called the 'greatest happiness principle.' An action is right (or wrong) to the extent that it increases (or decreases) the total happiness of the affected parties. Happiness is defined as advantage, benefit, good, or pleasure; and unhappiness as disadvantage, cost, evil, or pain. We must weigh the positive and negative impact of an act, or the morality of an action. The focus is on the consequences of the action. The calculations pertain to the person(s) affected by the act; the calculation includes a measure of the intensity, duration, certainty, propinquity, purity, and extent of the impact. This theory is quantitative, practical, and fairly comprehensive as it should consider all elements of a situation. A shortcoming is that, when performing the utilitarian calculations, it is not entirely clear who should be included, for what time period and duration. This can be a challenging task and requires much effort to solve every single moral decision. One approach is to develop 'rules of thumb;' for example, "it is wrong to lie." If a rule of thumb does not apply, then we can do all the calculations in detail to assess a unique situation or act. Contrary to Kantianism, Act U does not consider our sense of duty, only the consequences of an action. A weakness is that Act U cannot predict unintended consequences; it applies mainly when consequences are known and can be calculated. Act U is an objective, rational ethical theory that allows a person to explain why a particular act is right or wrong and is a workable ethical theory to evaluate moral problems for biomedical engineers [Quinn, M., 2005].

Rule Utilitarianism (Rule U): John Stuart Mill expanded on Bentham's Act Utilitarianism and developed the Rule U theory, which simplifies greatly the calculations required by using Act U to decide on a course of action. This approach supports the idea that if everyone adopts certain moral rules, the outcome will lead to the greatest increase in total happiness. It applies the Principle of Utility to moral rules, while Act Utilitarianism applies the Principle of Utility to individual moral actions. Both Rule U and Kantianism are focused on moral rules; some of the rules overlap, and both support the notion that rules should be followed without exception. However, the two theories derive moral rules in a different way. Rule U states that you follow a moral rule because its universal adoption would result in the greatest general happiness (consequences), whereas Kantianism states that you follow a moral rule because it is in accord with the categorical imperative (motivation and sense of duty). Performing the Rule U calculations is a simpler task than for Act U. One weakness is that utilitarianism forces us to use a single scale (units) to evaluate completely different kinds of consequences. Moreover, this approach ignores the problem of an uneven distribution of consequences. In summary, Rule U treats all persons as equals and provides its adherents with the ability to give the reasons why a particular action is right or wrong. It is a workable theory for evaluating moral problems facing engineers. We should act so that the greatest amount of good is produced and distribute the good as widely as possible [Andrews, G., 2005, Quinn, M., 2005].

Social Contract Theory (SC): Several philosophers developed the ideas around a social contract, but with very different viewpoints. Thomas Hobbes (1588–1679), in his book *Leviathan* (1651), advocated for absolute monarchy. Living in a time of war, he focused on the duty of individuals to support peace and also the right to defend themselves in situations of violence. Jean-Jacques Rousseau, for his part, advocated "collective sovereignty in the name of general will" in his book "*The Social Contract*" published in 1762.

John Locke (1632–1704), in his 1690 publication of *Two Treatises of Government*, advocated natural rights arising from being born. His work influenced many, including Rousseau, Voltaire, and the modern Charter of Human Rights adopted by the United Nations, and by several countries connected to this organization.

The Lockean concept of the Social Contract was invoked in the *United States Declaration of Independence*, and social contract notions have recently been included, although in a different sense, by thinkers such as John Rawls. Whereas SC is based on the benefits to the community and recognizes the harm that a concentration of wealth and power can cause, it is framed in the language of individual rights and explains why rational people act out of self-interest in the absence of common agreement. It also provides a clear ethical analysis of some important moral issues regarding the relationship between people and government. A weakness of the theory is that no one signed the social contract and some actions can be characterized in multiple ways. SC theory does not explain how to solve a moral problem when the analysis reveals conflicting rights; it may even turn out to be unjust for people who are incapable of upholding their side of the contract or for people who do not understand the rules [Quinn, M., 2005].

It seems that for our purpose of ethical decision-making for bioengineers, the most useful theory is John Locke's Rights Theory which states that everyone has rights arising simply from being born; the right to life, maximum individual liberty, and human dignity are all fundamental rights, and other rights arise as a consequence. The difference with Kantianism is that for Locke, duty is a consequence arising from personal rights. As mentioned above, the principles of Locke's theory are embedded in the 'Charter of Human Rights and Freedoms' in several countries, which guarantees to all citizens fundamental freedom of conscience, religion, thought, belief, opinion, expression, peaceful assembly, and association. The legal right to life, liberty, and security of the person and the right not to be deprived of these rights, except in accordance with principles of fundamental justice, are written in this law. This theory is usable in ethical decision-making by engineers, especially with regards to protection of the public and regarding the testing on human subjects [Andrews, G., 2005, Quinn, M., 2005].

Rawls' Theory of Justice : John Rawls (1921–2002), in his book *A Theory of Justice*, (1971), wrote that each person may claim a 'fully adequate' number of basic rights and liberties, such as freedom of thought and speech, freedom of association, the right to be safe from harm, and the right to own property, so long as these claims are consistent with everyone else having a claim to the same rights and liberties. Any social and economic inequalities must satisfy two conditions: First, they are associated with positions in society that everyone has a fair and equal opportunity to assume; and second, they are to be to the greatest benefit of the least advantaged members of society (the difference principle). In reality, regarding the first principle, we cannot imagine that every person will have the same skills, education, decisional power, and wealth; it is a certainty that some people will have more power and/or money than others. The second principle is more applicable; an example is that the poor should pay less tax than the rich. Like Locke, Rawls reaffirms the right to life, liberty, and security of the person, and the right not to be deprived of these rights except in accordance with principles of fundamental justice. This theory is applicable to decision-making by engineers [Quinn, M., 2005].

Moral dilemmas and issues change in time and place as new science discoveries are made and new technologies are developed. Examples are new research areas such as nanomaterials, stem cells, and the use of embryos. These are discussed in later chapters. However, certain universal principles apply to everyone and for all times, based on human rights and also on duty (categorical imperative). All human beings, of all races, ethnic backgrounds, gender, and sexual orientation are equal and must be given respect, dignity, and freedom from harassment and discrimination.

1.4   ETHICAL DECISION-MAKING PROCESS

The section below discusses the process by which bioengineers can come to a decision when facing a moral or ethical dilemma. Andrews describes the process of ethical decision-making as being similar, in its steps, to the engineering design process: recognizing the problem, gathering all relevant data, finding alternative solutions and the benefit and cost of each, selecting the best solution and optimizing it, and finally implementing it [Andrews, G., 2005]. It is critical to be aware of what is a moral dilemma that needs to be solved, and to acquire as much information about the issue or problem as possible, and identify possible solutions. In the analysis, two steps should be used: the first is to examine codes of ethics to determine which solution best respects the articles of the applicable code; a second step is to analyze each solution using ethical theories. The best solution would be the one that best fits codes of ethics and relevant ethical theories. In certain cases, one theory will be more applicable than another, and sometimes theories contradict each other. Through this analysis, we can come to see which of the solutions provide the most positive outcome (consequences). One guide to select the proper theory is to consider whether the ethical dilemma is a black and white situation (a categorical imperative) as for example when a safety issue can cause harm to the public. In that case, both Kantianism and the Rights theory can be applied. But if the issue is in a gray zone and has several potential solutions and consequences, then perhaps Rule U, or Act U, or Rawls' theory can be applied. The goal is to maximize the benefit and minimize the harm caused by the solution chosen, while respecting the principle of social justice.

Implementing the solution is the final step. If the problem threatens or can affect public safety, then this calls for immediate action to be taken. If the party responsible does not act, then one must escalate the ladder of authority to ensure that correcting actions are undertaken. If this does not work, then one needs to consider going public with the story if this can trigger some response towards solving the problem. Knowledge of ethical theories and the relevant codes of ethics helps engineers make the most ethical decisions when facing ethical dilemmas or problems.

1.5   CODES OF ETHICS

Bioengineers need to be familiar with several codes of ethics. Of course, the code of ethics of the professional engineering association to which the engineer belongs is foremost, as it guides the responsibilities and behavior of engineers in their work. Each state, province, and territory has its own Code of Ethics, which all have some common clauses, but also differ in minor ways. There also exist several other Codes of Ethics for engineers in the various technical and scientific societies; examples are codes established by the Institute of Electrical and Electronics Engineers [IEEE, 2011], the American Society of Mechanical Engineers [ASME, 2011], and the Biomedical Engineering Society [BMES, 2011].

Major grant agencies also produce guidelines that researchers and grantees must follow. The National Institutes of Health published *Ethical Guidelines and Regulations* at the URL: [NIH, 2011]. In Canada, the *Tri-Council Policy Statement: Ethical Conduct for Research Involving Humans* (TCPS) describes the policies of the Canadian Institutes of Health Research (CIHR), the Natural Sciences and Engineering Research Council of Canada (NSERC), and the Social Sciences and Humanities Research Council of Canada (SSHRC). These Agencies consider funding (or continued funding) only to individuals and institutions complying with the Policy on research involving human subjects. The policy was revised in December 2010 [Government of Canada, 2011].

Since bioengineers are likely to be testing devices, accessories such as sensors or transducers, or collecting samples or data from human subjects, then several additional codes of ethics are important to guide behavior for these studies. Although some of the codes are principally addressed to physicians, they also pertain to engineers who are involved in testing on humans. Principal codes are: The Hippocratic Oath, The Nuremberg Code, and the Declaration of Helsinki. The Hippocratic Oath is taken by physicians when they graduate; it sets out moral and ethical obligations to patients regarding the sanctity of human life, relief of suffering, to treat the ill to the best of one's ability, to preserve a patient's privacy and confidentiality.

The Nuremberg Code, created as a result of the Nuremberg trials at the end of World War II, deals with principles of human experimentation and the importance of obtaining voluntary and informed consent from subjects. Understanding the Hippocratic Oath and the Nuremberg Code is essential for any work involving human subjects. In addition, the World Medical Association (WMA) published the Declaration of Helsinki, a statement of ethical principles that provides guidance to physicians and others when performing research or testing on human subjects, human material like skin or tissue, and even when using de-indentified data. The Code was adopted by the 18th WMA General Assembly, Helsinki, Finland, in June 1964 and amended several times since then [WMA, 2004]. The latest version was adopted at the 59th *WMA* General Assembly, Seoul, Korea, in October 2008 [World Medical Association, 2004].

There is a call for biomedical societies to add clauses to their current codes of ethics that would integrate some of the principles of the Hypocratic Oath (which doctors adopt) and some aspects of the Nurenburg and of the American Medical Association's codes of ethics regarding experimentation involving humans. This involves the concept of ***Do No Harm*** and the first suggested application of this concept would be to discourage biomedical and biological engineers to participate in developing or working on the means to carry-out capital punishment. [Voigt and Ehrmann, 2010, Voigt, H., 2010]

The concept of ***Do No Harm*** can also be applied to many other biomedical and biological engineering endeavours. This is a very complex question which needs to be discussed and debated by professionals in these fields in order to add further guidelines in current Codes of Ethics to consider this principle seriously for current and future developments. This relates to being a socially responsible engineer.

Health Care Facilities, hospitals, universities, colleges, and research laboratories also have their own guidelines and ethical review process for all projects involving humans or animals to be carried-out within their walls. This includes course projects when students test or evaluate designs or experiments involving human subjects or animals. The Ethics Review Board of each institution has an application form to be completed, describing the project, the recruitment of volunteers, the inclusion and exclusion criteria for the selection of subjects to be tested; an application must also include an appropriate consent form, and other matters such as the storage of the data, and how to guarantee the privacy and anonymity of subjects when reporting results in a thesis or in an article. An Ethics Certificate is issued for each project with a timeline for which the certificate is valid. More details on the process of ethics applications are provided in the next chapter. The professional and positive interactions between engineers or scientists and physicians and mutual respect are important to ensure that design, testing, research and development of technology in a medical environment is successful.

1.6   OTHER ISSUES RELATED TO ETHICAL BEHAVIOR

Other key concepts to be aware of, in order to behave ethically, concern plagiarism, corruption, fraud, and being in a conflict of interest position without declaring this situation. It is also important to understand the concepts of intellectual property, confidentiality, privacy, and secrecy.

Plagiarism : According to the Merriam-Webster Online Dictionary, to plagiarize means to steal and pass off the ideas or words of another as one's own; to use another's production without crediting the source; to commit literary theft; to present as new and original an idea or product derived from an existing source. In other words, plagiarism is an act of fraud. It involves both stealing someone else's work and lying about it afterward [Budinger and Budinger, 2006]. In reference to students' work, the attempt to pass off the ideas, research, theories, or words of others as one's own is a serious academic offense. Copying an entire essay or a large section of a paper or book or buying a paper off the internet are serious cases of plagiarism. There are appropriate ways to use other people's texts; for example, quoting is for copying out the exact words of the original text and this requires that the entire text be in quotation marks or be indented and must include the source. For paraphrasing, the text is put in our own words and refers to the source of the information. Self-plagiarism refers to using text that we previously published. This can be done as long as we refer to the source of the original publication, and that the proportion of the earlier text reused is not substantial. A guideline is that twenty percent or less of the earlier work can be reused in a new text.

Corruption : "Corruption can be defined as the abuse of power, office, or resources by government officials or employees of organizations for personal gain, e.g., by extortion, soliciting or offering bribes" [Wikipedia, 2011c]. Several organizations deal with corruption, but the most cited source is the substantial report prepared by the Global Organization of Parliamentarians Against Corruption (GOPAC) entitled: *Controlling Corruption: A Parliamentarian's Handbook* (Third Edition). This report makes several recommendations and states that, to combat corruption, efforts must be continual and involve a constant maintenance of the institutions and systems of good governance nationally, regionally and globally [The Parliamentary Centre, 2000].

Misconduct : Each professional engineering association has a very detailed definition of misconduct which pertains to engineering works. In the case of experimental work or research, the Office of Research Integrity (ORI) in the USA defines misconduct as:

[F]abrication, falsification, or plagiarism in proposing, performing, or reviewing research, or in reporting research results. *Fabrication* is making up data or results and recording or reporting them. *Falsification* is manipulating research materials, equipment, or processes, or changing or omitting data or results such that the research is not accurately represented in the research record. *Plagiarism* is the appropriation of another person's ideas, processes, results, or words without giving appropriate credit. Research misconduct does not include honest error or differences of opinion [Office of Research Integrity, 2006].

Wikipedia defines a **fraud** as "an intentional deception made for personal gain or to damage another individual…Fraud is a crime and also a civil law violation. Defrauding people or entities of money or valuables is a common purpose of fraud, but there have also been fraudulent 'discoveries,' [for example] in science, to gain prestige rather than immediate monetary gain" [Wikipedia, 2011d].

There are several examples of scientists, physicians and engineers who committed a serious fraud in past decades. Two highly publicized cases are: Woo-Suk Hwang and Eric Poehlman. In an article referring to the Hwang case, Hand states "He (Hwang) is not alone; Newton, Sigmund Freud, and Nobel prizewinners have all been accused of being less than honest with data." Hand looked at fraud in science and found it not a rare occurrence. Hand quotes a paper in *Nature* reporting that, of 3247 anonymously surveyed scientists, 0.3% admitted having falsified research data at some point in their career, and 6% owned up to failing to present data if they contradicted their previous research [Hand, D., 2007]. Upon searching the term "fraud in science" on the internet, many cases can be found.

One can conclude that even though some cases become public, as the two cases presented in the Appendix show, many remain hidden as it is extremely difficult as a reviewer of a grant application or of a manuscript for publication to assess whether the data reported and the conclusions are fabricated or not, unless the fraudulent person is careless and provides data that do not make sense to the reader.

Intellectual Property (IP): is defined by the Canadian Intellectual Property Office (CIPO): "very broadly, means the legal rights that result from intellectual activity in the industrial, scientific, literary and artistic fields. IP rights, whether in the form of patent, trade-mark, copyrights, industrial designs, integrated circuit topographies, or plant breeders' rights reward this intellectual activity" [CIPO, 2011]. Wikipedia defines IP as "a legal entitlement which sometimes attaches to the expressed form of an idea, or to some other intangible subject matter. This legal entitlement generally enables its holder to exercise exclusive rights of use in relation to the subject matter of the IP" [Wikipedia, 2011f].

Copyright : refers to "creative and artistic works (books, movies, music, paintings, photographs, and software), giving a copyright holder the exclusive right to control reproduction or adaptation of such works for a certain period of time… A **patent** may be granted in relation to a new and useful invention, giving the patent holder an exclusive right to commercially exploit the invention for a certain period of time (typically 20 years from the filing date of a patent application)… A **trademark** is a distinctive sign which is used to distinguish the products or services... An **industrial design** refers to the form of appearance, style or design of an industrial object (spare parts, furniture or textiles, shapes)… A **trade secret** is an item of confidential information concerning the commercial practices or proprietary knowledge of a business" [Wikipedia, 2011i]. More information can also be found on the United States Patent and Trademark Office (USPTO) and (CIPO).

In addition to respecting the intellectual property of others, and protecting our own IP, there are other aspects of ethical behavior that are essential to understand. They are related to secrecy, privacy, and confidentiality. **Secrecy** refers to the keeping of secrets; information is withheld; a breach of secrecy is an unauthorized distribution of confidential information. In industry, it is common practice to have information that must not be divulged to competitors.

Privacy : refers to the freedom from intrusion or public attention; to be removed from public view or knowledge. There is an interesting article, *Bioinformatics and Privacy* that adds much information on this topic [Robison, W., 2010].

Confidentiality : refers to being entrusted with information that must not be divulged. This applies frequently in medicine and in research or testing with human subjects. It is usual that in the ethics application, a guarantee is made that the data will be kept confidential. More details are provided on this question in the next chapter. Confidentiality also applies to reviewers of manuscripts for journals or books, or grant applications; information found in these materials belongs to the person who submitted it and it is privileged information, so it must not be used by the reviewer except for parts that are already public information, like the references. The same applies also to reviewing a person's file for promotion or tenure or for a job or an award (see also Resnick and Shamoo [2003]).

Conflict of interest : "occurs when an individual or organization is involved in multiple interests, one of which could *possibly* corrupt the motivation for an act in the other" [Wikipedia, 2011b]. An example would be: An engineer is on a committee that will decide on the purchase of new electrodes for electrocardiogram machines. However, s/he was involved in the design of an electrode which is on the market and potentially one of the products to be considered. This is a conflict of interest if the engineer has shares in the company that sells the electrode or royalties based on sales. It is quite common for scientists or engineers to encounter situations like this. The important thing to do is to disclose this information to the committee and leave the room when this product is being discussed or chosen. Another frequent situation is when someone is on a grant or an award selection committee and has been a partner or relative of one of the persons being considered for a grant, prize or award. Again, the person has to disclose this fact and leave the room for this part of the discussions. A conflict of interest can be dealt with quite openly and easily with proper disclosure and removal of oneself for the decisions in these cases.

CHAPTER 2

Experiments with Human Subjects or Animals

2.1 EXPERIMENTS WITH HUMAN SUBJECTS

There has to be a good reason to involve humans in testing or in research. Shrader-Frechette argues that all professionals have a duty *not* to perform certain research (or experiments) that causes unjustified risk to people; that violates norms of free informed consent; that unjustly converts public resources to private profits; that seriously jeopardizes environmental welfare; or that is biased [Shrader-Frechette, K., 1994]. Consideration must be given first to whether a computer model can be used for the study. If this is not possible, we should consider if testing on animals would help the project.

When conducting research or experiments on human subjects, it is important to consider four basic concepts that enable to minimize harms and risks and maximize benefits; respect human dignity, privacy, and autonomy; take special precautions with vulnerable populations; and strive to distribute fairly the benefits and burdens of the research. These concepts are described below [Tong, R., 2007].

Beneficence : is the duty to do good for an individual and for all persons. It is important to ensure that the benefits expected to arise out of the testing or research outweigh the harm or risks to the participants or to society. Sometimes there are no direct benefits to the participants, but only benefits to people in the future. This is acceptable as long as the risks of participating are minimal [Tong, R., 2007].

Non-Maleficence : is the duty to cause no harm for an individual and for all persons. It is important to balance the benefits against potential harm. The concept should apply to physical, psychological, and spiritual harm. In cases where extreme survival measures are applied to patients, the decision should also consider quality of life issues [Tong, R., 2007].

Autonomy : This principle refers to the duty to maximize the individual's right to make his or her own decisions. Tong states "[N]ot only do autonomous persons have their own set of values and goals, they are also able to deliberate and reason about alternative courses of action, and to communicate their decisions to others" [Tong, R., 2007]. This assumes that patients have a right to refuse treatment and to be informed of medical consequences of their actions. For persons who are mentally disabled or are minors, parents or guardians must make the decision for them. However, in some cases, where the child or disabled person is able to comprehend what is to happen and the consequences of the choices, then they may also be asked to participate in the decision. An important point to make about autonomy is that the person who has to decide on a course of action needs all the information relevant to the decision in order to make an autonomous and informed choice. A paternalistic approach, where part of the information is withheld, removes the right to autonomy from the person making the decision. Autonomy is linked closely to the concept of 'informed consent,' as the latter can only be obtained when the persons are fully aware of all aspects concerning the testing or research project, its potential risks and benefits, and thus can decide if they agree to participate or not, or in a clinical situation, whether they agree to the test or therapy proposed.

Justice : is the duty to treat everyone fairly, distributing the risks and benefits equally to all persons concerned. Tong argues that "it is unjust to treat one person better or worse than another in similar circumstances unless there is a good reason for the differential in treatment" [Tong, R., 2007]. This may be a difficult decision when dealing with scarce medical supplies, or equipment such as artificial ventilators, where the demand may exceed the supply. Objective criteria are needed to make decisions such as removing a ventilator from one patient to place on another, or decide who will receive dialysis treatment when machines are in limited supply. The rationing of health care is posing some difficult ethical decisions to caregivers. Tong states that scarce resources must not be allocated according to social, moral, religious or economic condition, but rather on the basis of medical need. For example, will a limited supply of a drug or machine be provided for a financially-successful man, with a family, and church going, instead of an alcoholic single man or a divorced woman? In the past, hospital 'god committees' made the decisions for allocating dialysis [Tong, R., 2007].

Whenever the testing of new ideas or products involve humans, either as subjects, or when using samples or data collected from humans, an application for an ethics clearance certificate must be made to the institution's Ethics Review Board. In some situations, more than one certificate is needed. For example, if a student from a university is doing research with a hospital's data, the ethics application must be made to both the university where the student is enrolled and the hospital whose data or patients are studied.

2.2   APPLICATION PROCESS FOR ETHICS APPROVAL

Research Ethics Boards (REBs) exist in universities, hospitals, and all institutions involved in research or testing involving human subjects. Although the application process and the specific forms used may differ from one institution to another, there are some commonalities. In most cases, an application includes filling-in a form that identifies the principal applicants and any assistants involved in the study, the funding, and whether there are conflicts of interest to be aware of. An application normally requires a lay summary of the study, written in a way that is understandable by persons with a grade 8 education level. Many REBs require a proposal that includes the following sections: Background on why the research is needed and a short literature review on what else is being done on the topic;

the overall goal and objectives, the methodology, the selection criteria for the participants to be recruited and the inclusion and exclusion criteria for the selection of subjects. It should also include a statistical justification for the number of subjects to be tested or for the size of samples of blood, tissue, urine. For data, the size of the database to be collected or acquired must be justified.

The procedure for the recruitment of volunteers must be specified. No coercion of any kind must be used, in respect of the principle of autonomy, and it is important to avoid a situation of power; for example, a doctor should not be directly recruiting his/her patients; in this case, doctors need to have another person to do the recruitment. Similarly, a professor cannot recruit his/her own students; in this case, it is best for another professor or for students themselves to do the recruitment of peers. The advertisement must also be disclosed as an attachment to the application. If a questionnaire or survey is to be used, again this needs to be attached to the documentation.

The advertisement and recruitment letter must mention the following:

- the expected benefits or risks for the subjects; there are usually no direct benefits for the participants, but there may be some benefit for persons in the future; this must be explained;

- remuneration for the subjects; usually there should not be any, except in special circumstances, as this creates a strong incentive for poor persons to sign-up. However, it is quite acceptable to offer parking or transportation cost for people to participate. It is not usually a good idea to offer food or drinks in case people are allergic or in case the food may be contaminated. However, again, this depends on the experiment to be done. It may be acceptable to offer water in certain circumstances.

The following must also be explained:

- whether deception will be used in the experiment or not: this is unusual in engineering but could be used in some psychology experiments;

- anonymity of the participants and confidentiality; here it is important to mention how the anonymity will be preserved, either by assigning a number to the file, with no name except on the consent form, which must be kept separately, locked, and made accessible only to one or two key persons conducting the trial or testing; this is usually the principal investigator or the person who manages the recruitment and consent process;

- another point is whether pictures of the face will be taken or a video made; this makes the process more complicated as it can obviously identify the person;

- dissemination: if a thesis or papers are to be published, it is usually stated that only aggregated results will be included in such publications; if data from individuals are to be included, then it must not be identifiable to the individual in question;

- security of data: how the data will be stored and for how long must be specified;

- destruction of data: when and how this will be done; future use of the data must also be specified if this will be the case and a justification is needed; participants will be asked to give their consent if the data are to be used again in the future for a similar or for a different study. Even if this is specified at the time of the first study, it is possible that a new application may be requested for the later study.

The application usually contains several attachments; the information letter normally uses the 'you' pronoun and describes each step of the procedure, how long each step will take, and the benefits and risks, if any, to the subject. It also needs to address the remuneration if any is given for being a volunteer. Most institutions do not look positively on monetary rewards as this may prove to be coercive for poorer persons. As mentioned previously, travel costs and parking can be included. The most important part of the application is the Consent Form. Most institutions have a template for this attachment. It usually uses the pronoun 'I' and details again what is found in the information letter. One way to minimize repetition and provide a shorter document is to use what is called: Information letter and consent form.

- Informed Consent: The following is extracted from the university of Ottawa website, which follows the Canadian Tri-Council Policy Statement on Ethics in Research. Although the text below has some redundancies with the text in the previous section, it offers a good example of an institution's process and expectations. It adds some points not covered previously. The text on University of Ottawa's website follows:

"To give a free and informed consent, research participants must not be submitted to manipulation, undue influence or coercion. The protection of rights involves the following: The right of the participant to be informed of the specific nature and goal of the research so that he or she can grant or refuse consent in a free and informed manner; the right of the subject to be informed of harms and benefits of the research; the right to have one's private life and personal information protected; and the right for cultural groups to demand a respectful description of their heritage and their customs, as well as the discrete use of information about their life and aspirations.

The consent form must be written in clear, easily understandable and accessible language. It must include no less than the following information: Name and institutional link of any person seeking consent. Professors and students must indicate their affiliation with the university. Students must provide the principal investigator's or research supervisor's name, address and phone number, as well as this person's affiliation with the university.

The form must state that the research participants can ask the researcher any question about any part of the research being conducted. It must also state that information requests or complaints about the ethical conduct of the project can be addressed to the Protocol officer for ethics in research.

The name, the subject and goal of the research project, as well as the names of the agencies or individuals financing the research, when this information is relevant and when it may lead a person to refuse to take part in the research.

A clear, simple, non-technical explanation of what the participant's role involves, including: The general nature of the questions that will be asked, the tasks that must be accomplished and the observations or interventions that will be conducted; the length and frequency of the participant's participation; the location of the research work (add a description if the location is out of the ordinary, such as a sleep observation laboratory); the discomfort and probable harms created by the research. When the methods used will affect the private life of the participant or may have a negative or unpleasant effect, researchers must carefully explain the risk and discomfort when they ask for informed consent. Researchers must also mention sources of minor discomfort such as the application of probes to the skin as well as other harms (e.g., electrodes placed on the scalp will mess one's hair); the possible benefits that the investigator foresees. This statement must include a declaration about possible conflicts of interest, where relevant.

The rights of the research participant, including: The right of participants to withdraw from the project at any time, or during an interview, and the right of participants to refuse to answer questions without fear of reprisal or ill treatment (i.e., grades for courses); anonymity: The decision to publish data in a way that will allow participants to be identified must be soundly justified; confidentiality: When information gathered from participants cannot be kept confidential, subjects must be so informed; the rights of participants who are minors or who are vulnerable because they belong to a captive population (e.g., hospital patients, prisoners or students in a classroom) require particular care. A description of the compensation is to be provided, where applicable; publication of research results.

Providing a review or a summary of the research results is optional. A summary is compulsory if the research involves deception. Such projects must undergo an evaluation and lengthy discussion in compliance with the directives of the appropriate Research Ethics Board. The information the research participants need in order to file any complaint. The signature of the participant, parent, tutor or guardian, and a witness, must be included. The signature of the participant does not mean that he or she has given up any right, but rather that the participants has been informed of the requirements of the proposed research and that he or she agrees to take part in the research project. Researchers should obtain this signature for their own protection, especially as evidence in the event of legal action in which the participant claims that informed consent was not obtained" [University of Ottawa, 2011].

Research on children and infants : Some of the questions related to this topic are: Should research on children be allowed? Who should consent for research on children? How is research involving children different from research involving adults? How do we balance potential harm versus possible benefit in research involving children? To help answer these questions, the four basic ethical principles can be used: **Autonomy**: The duty to maximize the individual's right to make his or her own decisions. In this case, this can apply to parents, but not to a baby or young child. **Beneficence**: The duty to do good both individually and for all. If the outcome is uncertain, who should decide what is best for the baby or child? **Justice**: The duty to treat all fairly, distributing the risks and benefits equally; one consideration, for example, is that 90% of costs of newborn care is spent on babies who weigh less than 1000g. **Nonmaleficence**: The duty to cause no harm, both individually and for all. Some questions here could be: Is invasive neonatal intensive care acceptable if the outcome is uncertain? How does one balance the possibility of survival and the risk of poor long term quality of life? As for adults, one must have respect for human dignity and for free and informed consent by parents or guardians for their children or for vulnerable persons.

Children are entitled to special protection against abuse, exploitation or discrimination and are due respect for privacy and confidentiality, respect for justice, and for inclusiveness. Those who are not competent to consent for themselves shall not be automatically excluded from research that is potentially beneficial to them as individuals, or to the group they represent. If research on new drugs is not carried-out on children, then will they benefit from these? Or will physicians prescribe them based on results with adults, guessing the dose to be prescribed for babies or children? Moreover, the physiology of babies and children is quite different from that of adults and thus it is important to study the effect of various medications on children, provided this is done in the safest way possible.

Who should consent for children? Parents or guardians must be clearly aware of the research planned and of the risks and benefits to the child or infant. However, when the child has reached the age of reason, then it is advisable to obtain their consent in addition to the consent of the parents or guardians.

How to balance potential harm versus possible benefit in research involving children? Research may be acceptable if there is a potential benefit and a lesser potential for harm. Dr. Robin Walker, in lecture notes for an ethics course at Carleton University, provided examples:

1. A randomized clinical trial is to involve a substance which had been shown to benefit infants: The Research Ethics Board (REB) said: Yes, this is a well-designed clinical study with strong potential for benefit, limited risk and clinical equipoise (i.e. 'research' within Tri-Council guidelines in Canada). An example in neonatology is a clinical trial done to validate the usefulness of surfactant for treating respiratory distress syndrome in preterm infants, or a clinical trial for nitric oxide as a treatment for full term babies with conditions causing pulmonary hypertension.

2. Administration of a treatment for which there is limited evidence of benefit or safety, instead of a better known treatment: The REB said: No, you need to know that the

benefit outweighs the risk (i.e. this is 'experimentation', so not acceptable).In the last two years of the last decade, there has been consideration of using sulfadamil (better known as Viagra) for treating pulmonary hypertension in newborns. Although we know more now about this treatment, there was little evidence of either efficacy or safety during the time of the request to the REB as there was well studied alternatives such as nitric oxide already available.

3. Another situation is in extremis, with no alternatives and the safety information is known such as head and total body cooling after birth asphyxia: The REB said: Yes, if this is the baby's only chance, once the risk of ill effects is known and is definitely less than the potential benefit of using the technique (i.e. this is 'innovative therapy').

4. On the question of developing decision-support systems to help decision-making for parents of extremely low birth weight infants (ELBW); this project called PPADS (Physician-PArent Decision System) provides quick access to main clinical information to physicians and they can click on what parents should see about their infant regarding diagnosis, treatment plan, status of the patient, and decisions that need to be made such as "Full active care" or "Non-Escalation of Care" or "Withdrawal of Life Support and provide palliative care" or Do not resuscitate order": The REB said: Could the decision change as a result of the system? Could information provided lead to a 'wrong' decision? How would the parents be assisted through the decision-making process as they use the system? Would the research cause distress for parents asked to evaluate the system? The REB eventually gave full consent, including the parental involvement in the evaluation of the system. The system was tested with parents whose infant graduated from the NICU (neonatal intensive care unit) and the results were very positive [Frize & Weyand, 2011]. The next step is to test with parents whose infant died in the unit; a third study will involve any parents who will have some decisions to make, with a broader theme than for the former two evaluations.

Some things to consider are: The outcomes in the Neonatal Intensive Care Unit (NICU) may be uncertain and often there is not one single 'right' answer to 'ethical' decisions in this environment. One must remember that it is as important to 'do no harm' as to 'do good.' The clinical team tries to support parents in their decision-making for the baby, especially when the outcome is questionable or uncertain. Decisions can be made or changed at any point in a baby's life, as conditions or understanding of the situation changes. Development of decision-aids may be helpful for parents if designed with respect of all the ethical principles described above [Walker et al., 2009, Weyand et al., 2011].

The following section on research with children is extracted from an essay as part of the ethics course at Carleton by student Amanda Cherpak, with her permission.

"Research on children occurs much less frequently than research on adults due in part to strict ethical procedures and lower financial potential. This has lead to a current lack of knowledge of the effect of many pharmaceuticals on minors [Davidson and O'Brien, 2009]. In a review of paediatric drug labelling, it was found that 75% of drugs had insufficient testing in children and the same percentage of anaesthetics displayed no labelling for infants [Davidson and O'Brien, 2009]. The result is that there is an alarmingly high rate of off-license and off-label drug prescriptions for children since doctors are often left with no other therapeutic options [Pinxten et al., 2009]. The marketing of drugs tested for pediatric use has been discouraged by attempts to protect the ethical and legal rights of children by enforcing strict guidelines and constraints. This often requires testing that costs more than the probable financial gain and has left children with access to a lower standard of care than adults. Since a sick child is no less in need than a sick adult, this inequality violates the principle of justice as defined in Beauchamp and Childress [2009], and spotlights the need for a reform of the way research on children is promoted [Pinxten et al., 2009].

The strict guidelines mentioned involve navigating the sometimes grey areas of autonomy, beneficence and non-maleficence pertaining to research on children. The ethics of such research are not as clear as with research on adults since children do not always have the capacity to understand the consequences of the research or the ability to give informed consent. The principle of autonomy promotes self-regulation of what is done to one's body including consenting to research voluntarily and without coercion [Pinxten et al., 2009]. This can be achieved with able-minded adults provided they are given unbiased information that they are capable of understanding [Pinxten et al., 2009]. Minors are not legally capable of making autonomous decisions and consenting to research; however, their reasoning and decision-making skills can be widely varied and in some cases should be considered [Davidson and O'Brien, 2009].

An infant may have no capacity to understand or consent to research; however, older children may be able to form their own opinions and choices that should not be ignored. One of the motivations for consenting to research is the ideal of altruism, which leads one to accept possible risks and burdens for the benefit of others; however, these morals do not develop until around 11–14 years of age [Eisenberg et al., 1991]. Children are also more vulnerable to coercion by their parents and by researchers, which has led to closer scrutiny of ethics proposals for research on children than for similar cases involving adults. The compromise currently in use is to require the parent's fully informed and free consent as a proxy in all cases and, with the discretion of health-care workers, the assent of the child. This involves the child agreeing to the research with the implied limitations in understanding the full scope of their decision [Davidson and O'Brien, 2009]. Ethical guidelines often do not state a specific age at which assent should be obtained, since the maturity of children and the complexity of research varies widely; however, certain organizations have adopted age 7 as a minimal age [Davidson and O'Brien, 2009]. A study by Ondrusek et al. [1998] alternatively suggested age 9, since it was found that understanding of the research was poor in younger children and in many cases the children believed that failure to participate would displease others. This highlights the need for clear explanations and the use of plain language to ensure that the child does not receive the wrong information.

This method of consent is not without flaws, as it raises the question of the appropriate level of risk parents can consent to on behalf of their child. Since it has already been established that younger children do not have the capacity to act altruistically, it also questions the use of placebos. Many studies require the use of a placebo as a baseline comparison for the proposed treatment; however, if another method of treatment is readily available, withholding that treatment can put the patient at an unnecessary risk, which is in violation of the principles of beneficence and non-maleficence [Davidson and O'Brien, 2009]. When the research participants are adults, the principle of autonomy can outweigh this violation as long as the risk is as low as possible. According to Tong, R. [2007], this compromise is allowed as long as there are better reasons for acting on the overriding norm (autonomy) than on the infringed norm (beneficence and non-maleficence).

The situation is different with children since the ethics of research with no net risk to the child are not clear [Davidson and O'Brien, 2009]. The result is that the acceptable level of risk for research on children is lower than for adults, but considering that the child may benefit in the future, a minimal risk is tolerable [Pinxten et al., 2009]. Minimal risk is defined as the level of risk associated with everyday life, and if the risk of the research is above this threshold, it must be weighed against the possibility of direct benefit to the child. If there is no direct benefit, then the research must aim to provide a benefit for children of a similar age and condition [Pinxten et al., 2009].

As described above, there are several additional considerations when clinical research is conducted on children. Several government bodies have recently introduced new initiatives to encourage the expansion of our knowledge of treatments for children; however, financial incentives must be closely monitored to avoid sub standard research [Davidson and O'Brien, 2009]. Research on children should be allowed; however, it must be handled with great care and the health and well-being of the child should be protected at all times [Cherpak, A., 2009]."

2.3   TESTING ON ANIMALS

As with humans, there must be good reasons to use animals for research. Again, it is important to assess whether computer models or cell cultures can help answer the research question as a preliminary step prior to deciding to use animals. Several questions need to be asked: How can research results derived from animal testing be extrapolated to humans? What assurances exist that stolen or lost pets are not used in research? Why is it important to conduct product safety tests on animals when 'cruelty-free' products are available? Are the animals in laboratories suffering and in pain? What happens to animals once an experiment is completed? Why are increasing numbers of animals used in research? Do we really have the right to experiment on animals? What about their rights? It is recommended to become familiar with guidelines on the treatment of laboratory animals in the country where the research is being planned. (Other sources: Budinger and Budinger [2006], NIH [2011], Resnick and Shamoo [2003], Shrader-Frechette, K. [1994].)

2.3.1   HARM VERSUS BENEFITS

Some argue that research on animals is needed to advance medical cures and basic understanding of nature. Others claim it involves suffering or the sacrifice of animals and the knowledge gained does not justify the harm. Research on animals is growing around the world because of the increased use of transgenic animal models to study aspects of gene function, expression, and regulation. Animal models of human genetic diseases allow the study of treatment possibilities, such as gene therapy [Budinger and Budinger, 2006]. Some argue that transgenic animals are in a different community of ethics since they would not have been bred and would not exist if not for their purpose as research tools. Opponents argue that since the harm of animal research is definite, and the benefit is only probable, it is unjustifiable. Similarly, cloning animals has a success rate of less than fifty percent, so many animals are born dead or malformed. Furthermore, the creatures that suffer will not benefit from their suffering. On the other hand, many life-saving procedures have been developed through animal experimentation. Thus, to balance the pros and cons of animal research, there are strict regulations which ensure adequate justification and minimal suffering for research on animals [Budinger and Budinger, 2006].

When considering the use of animals for research, a rationale must be presented on the appropriateness of the choice of the species, and the number of animals to be used must be justified using a statistical argument. A complete description of the exact protocol to be used must be provided. Procedures must be designed to ensure that discomfort and injury to animals are limited to what is unavoidable in the conduct of scientifically valuable research, and that analgesic, anesthetic, and tranquilizing drugs will be used where indicated and appropriate to minimize discomfort and pain. The euthanasia method to be used must be specified [Budinger and Budinger, 2006]. There are means to reduce the need for animals. A solution to the conflict between animal ethics and medical research needs could be to substitute new technologies for animal experiments. For example, the use of microarrays, new imaging methods, and the use of selected transgenic animals can help to replace animal studies [Budinger and Budinger, 2006]. If research on animals is justified, then an application to the appropriate committee on animal care of the institution must be made and strict guidelines adhered to in designing the experimental protocol.

CHAPTER 3

Examples of Ethical Dilemmas in Biomedical Research

Several websites provide a list of potential ethical dilemmas in medical and biomedical research. One example is presented in this chapter which has interesting aspects regarding ethical issues: reproductive technologies. This term includes various techniques; some pose little ethical problems, but others raise serious ethical concerns.

3.1   REPRODUCTIVE TECHNOLOGY

The term reproductive technology (RT) includes several types of biomedical interventions that can help a woman when she considers to have or not to have a child. RTs include: artificial insemination (AI), *in vitro* fertilization (IVF), embryo transfer and freezing, gamete intrafallopian transfer (GIFT), zygote intrafallopian transfer (ZIFT), intracytoplasmic sperm injection (ICSI) and assisted hatching—embryo micromanipulation [Health Policy Coach, 2011].

Other technologies are used for testing how the foetus is developing and whether abnormalities are detected or not. These include: ultrasound, chorionic villus sampling, aminiocentesis and laparoscopy. These aspects are discussed in a section below. Other technologies are more controversial and raise important ethical concerns. These include: sex pre-selection, designer babies, genetic screening, and cloning.

The chapter begins with brief explanations of the various artificial fertilization techniques, followed by descriptions of the some of the testing technologies. The next section provides details on some of the more controversial aspects of RTs. The next section provides a brief summary of a Canadian study on what a sample of the Canadian population think about these technologies. The chapter concludes with a discussion of issues raised by body enhancement technologies.

RTs provide artificial fertilization procedures for women and men who are not able to have a child. Specialized infertility treatments are designed to increase the number of eggs and/or sperm, or bring them closer together, resulting in an improved probability of conception not otherwise possible. Collectively, these medical procedures are called assisted reproductive technologies (ART) The principal methods are listed briefly: Artificial insemination (AI) involves injecting sperm through a narrow catheter into a woman's reproductive tract. AI can be done with either the husband's or with a donor's sperm. In vitro fertilization (IVF) is a procedure that involves retrieving eggs and sperm from the bodies of the male and female, placing them in a dish, then the fertilized eggs are transferred into the female uterus where implantation and embryo development will hopefully occur as in a normal pregnancy. In Gamete Intra Fallopian Transfer (GIFT), fertilization occurs naturally within the female body instead of the laboratory dish. Intracytoplasmic Sperm Injection (ICSI) involves the insertion of a single sperm directly into the *cytoplasm* of a mature egg, or *oocyte*, using a microinjection pipette, or thin glass needle.

Some questions regarding RTs are: What is the moral status before sperm and egg come together? What about the moral status after sperm and egg come together but before implantation in the uterine wall? What is the moral status after implantation? What about left-over frozen embryos? Typically, during fertility treatments, several fertilized eggs, embryos, are stored as part of their treatment. These are kept frozen for unanticipated future need. Many major fertility centers have thousands of these. There does not seem to be a federal agency overseeing this. Sometimes embryos are donated to infertile couples, but legal issues about parental rights persist. What are the individual, moral constraints of trying for reproduction 'at any cost'? What are the dehumanizing aspects in using these technologies? ARTs have medicalized the process of having a child which normally would arise from an 'act of love.' The question also arises of selective abortion with multiple fetuses, as multiple births frequently result when using ARTs. What about the pregnancy of older women? What about the genetic testing for selecting the fetus to be implanted into the womb? Several important social issues arise from these choices such as the occurrence of vast demographic changes, a shift in male/female ratio, a shift in age of parents relative to their children, homogeneity in society, and possibly a decline of two parent families. (For more information: see Hinman, L. [2011], University of San Diego.)

Another issue is the question of surrogacy, which occurs when a woman carries to term the fetus for another person. That fetus may be from the egg and sperm of the couple who want to raise the child or it may be donor eggs or donor sperm. What are the rights of the surrogate mother? Of the individuals who ask her to be a surrogate? There are various possibilities: an *intentional mother*, the woman who wants to have the child; an *intentional father*, the man who wants to have the child; the *genetic mother*, the woman who supplies the egg for the embryo; the *genetic father*, the man who supplies the sperm for the embryo; the *gestational mother*, the woman who carries the embryo to term and gives birth to it; the *nurturing mother*, the woman who raises and nurtures the child from infancy as her own; and the *nurturing father*, the man who raises and nurtures the child from infancy as his own. This can lead to difficult situations and several have ended up in court. For example, the surrogate mother may change her mind about giving up the baby when it is born, especially if the egg came from her body. (For more details, see McNeil et al. [1990].)

A study regarding attitudes to new RTs was carried-out by the Royal Commission on New Reproductive Technologies in Canada in 1993. The study tested the views of three separate groups of Canadians on: 1. the scientific developments and the technology around ARTs; 2. the family and children; 3. assisted conception methods and practices; 4. accessibility; 5. alternatives if they were faced with infertility; and 6. who should pay for these services. Additionally they were asked what they thought of prenatal diagnosis, fetal tissue research, and who should decide what policy will govern these matters A brief summary of the results of the study with one of the groups is provided below:

On the question of scientific developments, over 2/3 welcomed these; 10% expressed fear towards these; 21% indicated some fear; 35% worried that medical science was moving too fast for society to control its uses; 44% were not worried about the pace of medical science. On the medical technologies themselves: 48% were not worried about their safety, but 18% viewed them with skepticism; 1/3 of participants were unsure about their safety; 70% supported the existence of technologies to assist people who have difficulties in having children. For complete details on the study, consult the publication by the Royal Commission on New Reproductive Technologies [1993].

3.2 TECHNOLOGIES USED FOR TESTING THE STATUS OF THE FETUS IN-UTERO

These include chorionic villus sampling, laparoscopy and amniocentesis, and ultrasound. These technologies are used to detect if the fetus is developing normally and to identify abnormalities at the earliest possible stage of the pregnancy. However, some of the technologies can also be used in negative ways, which is discussed later in this chapter. A brief definition of the terms listed above follows.

In the procedure called amniocentesis, a small amount of amniotic fluid containing fetal tissues is extracted from the amniotic sac surrounding a developing fetus. The DNA of the tissues is examined for genetic abnormalities. Amniocentesis is usually performed in the second trimester of pregnancy. Amniocentesis can be performed as early as 11 weeks, but is considered too risky by most providers.

Chorionic villus sampling is typically performed in the first trimester of pregnancy. Fetal cells are obtained through biopsy of the cells that will become the placenta to determine all disorders that can be diagnosed by amniocentesis except the presence of neural tube defects [Reproductive Health Technologies Project, 2011].

Ultrasound is routinely used to determine fetal viability, the number of fetuses present, the position of the fetus and to estimate fetal age. Sex, fetal structures, and some fetal malformation can also be determined using ultrasound, depending on age and position of the fetus and extent of the examination [Reproductive Health Technologies Project, 2011].

3.3 TECHNOLOGIES THAT RAISE ETHICAL CONCERNS

The development of fetal ultrasound monitoring had, as primary function and goal, to help assess the status of the fetus in-utero during the pregnancy. It was non-invasive, easy to use, and provided information on the development of the fetus. Routinely done around three months gestation, the images allow to see if the development of the fetus appears normal or whether serious defects can be detected.

The College of Physicians and Surgeons of Ontario state:

Physicians must ensure that all diagnostic fetal ultrasounds are ordered and conducted for appropriate clinical indications, in accordance with relevant statements and guidelines. The purpose of an imaging examination should always be to obtain information relevant to the diagnosis or treatment of a patient. Therefore, when ordering the diagnostic fetal ultrasound, the physician should specify the clinical indications. If a physician orders or performs a diagnostic fetal ultrasound for medical reasons, they may provide their patients with any picture or video of the fetus that is created as a result of that imaging examination. However, it is inappropriate and contrary to good medical practice to use ultrasound only to view the fetus to obtain a picture or video of the fetus or to determine the gender of the fetus [CPSO, 2010].

However, in countries such as China and India, and several others, the fetal ultrasound exists in small clinics in many cities to allow parents to discover the sex of the fetus. One consequence in these countries, where males are more valued than females, is that millions of female fetuses are aborted after the test confirmed their sex. The societal consequence of this is that millions of women are missing on the planet. This is especially true in China where, in some rural areas, there are 130 men for 100 women. Gender balance cannot be sustained and some men, in China for example, buy girls or young women to marry them; some actually resort to kidnapping to find a wife [Hvistendahl, M., 2011].

Mara Hvistendahl has also studied the situation in India and reports that amniocentesis tests and ultrasound scans have led to more than 160 million girls being aborted in Asia since the 1970s. Today, 112 boys are born for every 100 girls in India, against the natural sex ratio at birth of 105 boys for every 100 girls. In Delhi, there are only 836 girls under 7 for 1000 boys of same age. Between 2001 and 2011, the ratio of girls to boys worsened in seventeen states in India, in spite of the new law forbidding the testing for gender passed in 1994 [Hvistendahl, M., 2011].

Laws in the countries misusing this technology to abort female fetuses have only recently been enacted. For example, determining the sex of a fetus is illegal in India since 1994. Persons who opt for it and doctors who do the test and reveal data are held guilty under the Prenatal Sex Determination (Prohibition and Regulation) Act. The act aims to prevent female feticide, which, according to the Indian Ministry of Health and Family Welfare, "has its roots in India's long history of strong patriarchal influence in all spheres of life". A story in July 2011 reports that a gynecologist in Saki Naka, India, was caught determining the sex of a fetus of a pregnant woman [Planet Powai, 2011].

Engineers cannot work alone on these issues. They need to work with lawyers and governments to ensure a proper use of their designs and inventions.

Genetic testing can also have both positive and negative consequences depending on how and why it is used. On the positive side, genetic testing can help detect serious defects or diseases, and if done early enough during the pregnancy, it can allow parents to have a choice as to whether they want the fetus to be aborted or grow to term. Of course the values and beliefs of the parents play an important part in this decision. The ethical questions are: What is a serious defect or disease that would warrant an abortion? It should not be the color of the eyes or the sex of the baby. On the other hand, the probability that the fetus will have Huntington's disease, for example, or severe brain damage, may provide substantive reasons to weigh all possible decisions and their consequences. Parents have to provide the long-term care for a child with a genetic disorder. Huntington's disease causes degeneration of cells in certain areas of the brain which causes uncontrolled movements, loss of intellectual faculties, and emotional disturbance. It has no known cure and the medical costs for treatment are very high.

Another approach referred to as 'Designer Babies' uses IVF and genetic testing to select the fetus that is free from serious diseases. But a serious ethical concern is that some clinics specifically do this in order to select the fetus that has the desired characteristics such as color of eyes, hair, sex, etc. If this technique was practiced by many, it would be easy to imagine that genetic homogeneity in the world would become a reality.

Another topic which leads to many debates is the use of stem cells for research and clinical treatments. The controversy has more to do with the source of embryonic cells than on their use. Stem cells can be obtained either from embryos or from adult cells collected for cloning. In the case of the latter, the cells can only replicate themselves, thus for the same function which they served in the adult. However, embryonic stem cells can differentiate into many types of cells, so they are more useful for the treatment of diseases where organs or other types of cells need to be replaced. Stem cell research has applications in the treatment of diabetes, Parkinson's, heart disease, liver disease, arthritis, paralysis, and many other serious medical conditions.

Embryonic stem cells can be obtained from aborted embryos, from embryos remaining after IVF, from embryos created for research by IVF techniques, or by therapeutic cloning. Many embryos are left over after IVF and most are destined to be destroyed after a certain period. The family undergoing IVF would have one embryo implanted in the uterus, and if this is successful, then all others may be destined to freezing or destruction. The debate is still raging about abortion, so again, using embryos from this source remains controversial [Resnick and Shamoo, 2003].

3.3.1   ETHICAL QUESTIONS ON EMBRYONIC STEM CELL RESEARCH AND THERAPY

This topic raises more questions than we have answers, as shown below: For example, is therapeutic cloning (expanding embryonic stem cells in culture and using them for restorative applications) ethical? Is it morally acceptable to use donated embryos with a potential for life to obtain stem cells? Stem cells can be taken from a blastocyst grown from a blastomere; obtaining the blastomere does not sacrifice the embryo. Is this ethical? Is it morally acceptable for parents to donate their spare embryos for research purposes? Is it ethical to create or use an already formed embryo for the purpose of performing experimental research? What about using a frozen embryo that has died following thawing? It is now possible to do pre-implantation screening and to select IVF-conceived embryos to ensure tissue compatibility with an ill sibling needing stem cell transplantation. Is this ethical? If it is shown that adult stem cells can be substituted for embryonic stem cells, is it ethical to proceed with embryonic stem cell research? Are the methods of the acquisition of human eggs for therapeutic cloning unethical? Is it ethical for the U.S. government to limit research that has the potential to benefit millions of people with serious illnesses? Should stem cell acquisition procedures be regulated by the state or the federal government, including the FDA? Is it appropriate to limit the number of stem cell lines available for research? Is it ethical to obtain embryonic stem cells from an embryo conceived for the purpose of obtaining stem cells? Is it ethical to obtain stem cells from a spontaneously aborted fetus? What should be society's role in promoting or negating stem cell research? These questions show that many ethical concerns are raised in this field of research and should be debated by researchers prior to deciding on a course of action. This also applies to persons (parents and patients) who wish to benefit from such technologies and approaches; all aspects and their consequences need to be well understood before deciding on their use.

Discussions on these issues can be found on several websites and in books. The important thing to remember is that the ethical theories and codes of ethics can help to decide how to think about these issues and what path to take if a decision on their use is needed.

3.3.2   BODY ENHANCEMENT TECHNOLOGIES AND ETHICS

This chapter ends with a discussion on technologies that enhance a body's function or capability, raising again several ethical questions. This topic is well known in sports where drugs are sometimes used to enhance performance. At what point does drug use make the playing field uneven for participants? No matter what the morality of taking drugs might be, is there an overriding duty to follow the rules of the sport with respect to doping? Are there moral reasons that drug use in sports should be forbidden? Does prohibiting drug use restrict the autonomy of the athlete who wishes to use drugs? In response to the increasingly sophisticated testing in the competitive world of sports, a market is developing for drug free urine [Budinger and Budinger, 2006].

Drugs can be used to change behavior, as in the case of patients with attention deficit disorder (ADD). However, these drugs have been shown to have some serious secondary effects. At what point should a family decide to use the medication? Is it to make a normally boisterous child more docile or for a serious case of hyperactivity?

There are many other types of body enhancements that are less well known but are important to discuss. Enhancement can mean an intended change to improve an already normal, average individual's features by surgery or drugs, or may mean a corrective action needed for a serious medical condition. A related question is: What is normal and what is defined as enhancement? Are reading glasses an enhancement or has this become normal with ubiquitous use? Buchanan writes:

> Taken together, literacy and numeracy are profound and far-reaching cognitive enhance-ments. Computers, building on the platform of literacy and numeracy, extend human cognitive capacities even farther…We *now* consider literacy, the use of computers, and the ability to engage in large-scale coordinated, complex activities through the function-ing of institutions to be "normal" capacities of human beings, but for most of the time during which human beings have existed they were not [Buchanan, A., 2008, 4].

Hogle mentions the President's Council on Bioethics report of 2003 entitled: *Beyond Therapy: Biotechnology and the pursuit of Happiness* that defines enhancement: [http://www.bioethics.gov/reports/beyondtherapy; note the report is still available but the Council was disbanded in June 2009.]

Enhancement technologies are most commonly defined as interventions intended to improve human function or characteristics beyond what is necessary to sustain health or repair the body. The difficulties inherent in attempting to analyze enhancements can be seen in this definition itself. What is necessary to sustain health? At which point does a repair become something more than restorative, and for which (and whose) purposes are interventions defined as "therapeutic"? (Hogle, L. [2005, 696–7], from Presidential Council for the Study of Bioethics [2003]).

Hogle proposes to assess normalcy through the statistical concept of a normal distribution of a population, involving human traits or features, and to consider outliers as being 'less than normal' or being 'enhanced or augmented.' Hogle references Canguilhem, G. [1989] who had suggested that:

[I]n health, the idea of a normal curve was consistent with conceptualizing disease in terms of a continuum of qualities that constitute health states.

Because of the state's interest in using information to manage populations' health and labor, however, distinct categories were constructed, creating a dichotomy of normal and pathological states (Hogle, L. [2005, 696–7], from Canguilhem, G. [1989]).

Hogle refers to an article by Davis who argued that "classifications and definitions can then be used by governmental or medical authorities to create guidelines for how to deal with things and people outside the norms" (Davis, L. [2002, 107], from Hogle, L. [2005, 698]). This makes sense when we look at the guidelines on which diagnosis-related groups (DRGs) are decided. [In a DRG prospective payment system, Medicare pays hospitals a flat rate per case for inpatient hospital care so that efficient hospitals are rewarded for their efficiency and inefficient hospitals have an incentive to become more efficient;… a rate of payment is based on the "average" cost to deliver care (bundled services) to a patient with a particular disease. http://oig.hhs.gov/oei/reports/oei-09-00-00200.pdf]

It can be assumed that this type of reasoning will be used when enhancement technologies are to be considered for the general population, especially in cases where there is popular and/or political pressure to approve these under a Medicare plan. Of course, the rich will always find ways to get what they wish by paying for the services wherever they can be obtained. In this latter case, the offer of these technologies will depend to some degree on national policy and regulations of devices instituted by organizations like the Federal Drugs Administration (FDA) in the USA and the Medical Device Bureau in Canada.

One way to think about the ethics of these developments is to use the ethical theory of utilitarianism. For each new enhancement created or invented, one can think: Will this produce more happiness to more people or more happiness to a few people and unfairness (here considered as unhappiness) to more people? Let us look at a few examples that are contemporary. Cosmetic surgery can be performed after an accident or in a case of severe burns; corrective surgery can help the individual to recover as normal a life as possible, especially after severe disfiguration. What about surgery without medical need? Take, for example, the 'Biojewellery' project in the UK, funded by a public granting agency; this project attempted to raise awareness and increase public comfort with the idea of bioengineering bones. They took bone samples from couples interested in having wedding rings grown from their own bone tissue.

> The couple's cells were grown at Guy's Hospital (London), and the final bone tissue was taken to a studio at the Royal College of Art to be made into a pair of rings. The bone was combined with traditional precious metals so that each partner has a ring made with the tissue of their partner" [Biojewellery, 2011].

Needless to say this project raised a lively ethical debate in the media and in scientific circles. Miah, in an article in the newspaper *The Guardian* writes:

> Nowadays, we can lengthen our legs, chemically enhance our mental ability and perhaps even genetically modify ourselves to become stronger, faster or more resilient to wear and tear… The kinds of enhancements we must seek for humanity should not lead us towards a world where we all aspire to look the same as each other, which is a criticism often leveled at the cosmetic surgery industry. Rather, we should encourage human enhancements that amplify human variation. That's what I expect from human enhancement technologies and this is what humanity excels in, as the history of fashion reveals [Miah, A., 2009].

Looking at changes in body parts, two examples are cited: The dancer Cynthia Hess has successfully claimed the cost of her silicon breast implants as an allowable business expense on the basis that the operation had been a "marketing decision, akin to the cost of industrial retooling, which transformed her business prospects" [Wikipedia, 2011m]. Perhaps the most remarkable instance of the transmutation of the body, combining surgical, prosthetic and computer technology with consumerism, is that of the French professor and performance artist Orlan. Since 1990 she has altered her face and body through a series of performance art operations guided by a computer generated image to which her face is recut. She markets photographs and film of the surgical performances, as well as preserved body parts, complete with a label saying "This is my body, this is my software." 'Orlan' is not her name. Her face is not her face. Soon her body will not be her body [Wikipedia, 2011h].

There are medical imperatives as in the use of prostheses, the implantation of pacemakers, defibrillators, and stimulators. The philosophical question is: When does a human become more of a machine than a human? Do technologies create an unfair advantage? Or does it provide equal opportunity for persons who suffer from some disability? Here we look at the example of Oscar Pistorius, a South African man who was a double amputee at the age of 11 months. Pistorius wears prosthesis on both legs; he is called "the Blade Runner." He became Paralympic World Championship in 2006, placed first for the 100m, 200m and 400m men's events, and broke his 200m record. He requested to compete in the 2008 Beijing Olympics in the regular events, but he missed qualifying by 0.7 seconds and so was not selected by the South African team for the 4X400 meter relay. However, he will be part of the South African team for the 2012 Olympics in London [Wikipedia, 2011j].

One issue that arises from the use of body enhancement technologies is related partly to the purpose of the procedure. For example, does the change bring unjust advantages to the individual? If the modification results from unnatural genetic manipulations, then will this forever be a part of the DNA and passed down to future generations? It remains to be seen whether the emerging technologies will outstrip our ability to use them wisely and justly. Does society have a duty to institute policies to limit enhancements that carry potential risks to individuals or to societal norms? The authors, T.F. Budinger and M.D. Budinger pose the following questions:

> Will the competitive advantage of improvement in physical appearance (ex: cosmetic surgery), behavior modification (ex: Prozac), and intellectual performance (ex: drugs or genetic change of the future) be unfair if the enhancements are not available to all? If enhancements in education or athletic training are imposed upon a child by a parent, is there an ethical issue regarding the child's autonomy? [Budinger and Budinger, 2006, 405].

Ethical issues involved in using drugs to enhance performance in sports include the duty to obey the rules of the sport, not putting other athletes at risk by pressuring them into taking drugs so that they can compete as well; everyone must remain conscious of the virtue of competing using natural capabilities.

Using drugs for behavioral and physical enhancements raises other ethical issues. A student using stimulants before writing a test may put other students at a disadvantage, but the one using drugs also puts his/her own health at risk. What are the moral issues regarding the use of Prozac for individuals who are not diagnosed with clinical depression? Cosmetic surgery is considered unethical by some because those who need surgery for more serious conditions and cannot afford it may be less likely to receive it than those who can pay for surgical enhancements from surgeons in private practice. One may need cosmetic surgery to increase self-confidence and acceptance of oneself; in this case, the need for surgery can be considered just as important as surgery needed for medical reasons.

3.3.3   ADVANCEMENTS ON REGENERATION OF BODY PARTS

Research on tissue engineering and on artificial organs is making huge steps. The field began with skin substitutes in the 1980s and 1990s. Badylak and Nerem note:

> By the mid-1990s the emphasis on tissue replacement with ex-vivo manufactured products had evolved to include broad strategies to induce both in-vivo constructive remodeling of cell-based and cell-free scaffold materials and true tissue regeneration, marking

the emergence of the era of regenerative medicine. The desire to construct tissues and organs more complex than partial-thickness skin substitutes made it obvious that the simple self-assembly or directed assembly of different cell types ex-vivo would be inadequate to meet all the challenges. A critical issue for successful regenerative medicine applications is the source from which the cells are harvested [Badylak and Nerem, 2010].

Since then, other researchers have also made progress on regenerating blood vessels. Taylor, D. [2011] and her colleagues at the University of Minnesota "blends research using stem cells, genes, and devices to develop novel cardiac and vascular technologies—ones to prevent, treat, and hopefully one day, cure heart ailments." Dr. Taylor is currently involved in both laboratory and clinical studies using cell therapy to treat disease. On Dr. Taylor's website, we can see a video of a beating heart of a rat created by regeneration. These advances are very exciting, but they also raise ethical questions that will need to be addressed before the outcomes of these research projects become a reality for humans.

An article that complements the information presented in this chapter is: *Ethics of Medicine, Biology and Bioengineering at the New Critical Crossroads for Our Species—Beyond Aristotle and Hippocrates;* the author discusses genetics, biomachines, and synthetic biology [Bugliarello, G., 2010].

CHAPTER 4

Technology and Society

An important consideration when designing new devices or systems is the assessment of their potential impact on society. New developments, particularly in robotics, genetic engineering, and nanotechnology can raise serious concerns. The Center for the Study of Technology and Society summarizes the issue:

> The same technologies that will let us cure diseases, expand the economy, and overcome every day inconveniences can theoretically bring about catastrophes. Is the risk of an apocalypse serious enough for us to relinquish the current pace of technological innovation? [Specht, J., 2000].

Technology is often ahead of societal guidelines and laws. Are we living in an era where technological determinism prevails? Wikipedia defines technological determinism as a reductionist theory that presumes that a society's technology drives the development of its social structure and cultural values [Wikipedia, 2011n]. A website presents interesting views on technology; it states:

> Some who would argue that technology is autonomous technology usually see it as the cause of one problem or another. Yet, different people see technology as causing different problems. Since World War II scientists have seen technology as a moral dilemma where their work can have profound effects on the human race and on the planet. Sociologists see the problem as the increasing complexity and rate of change which technology is bringing about in society; technological changes, they argue, outpace the ability of individuals and societies to adapt. To others, technology is seen as a dominating force over society, posing a threat to human freedom [The UK Technology Education Centre, 2011].

Can society provide sufficient controls and moral and ethical guidance to prevent irreparable harm to our world and to the people in it? Engineers must be able to understand how to verify the models that assess the impact, both positive and negative, of technological development on society and need to be aware of universal responsibility and inter-dependability of engineers and society. This is particularly true for biomedical engineers whose work frequently impacts people and health care. One must remember that there is always a dark side to how technologies will be used once they are created, even though the benefits of the new development may be foremost in the mind of the designers.

Discussing the impact of technology on society and on people is a component of the expected ethical behavior of engineers. Technological development brings benefits and hazards in its wake [Joy, B., 2000]. Engineers can ensure that a positive impact on society predominates and efforts should help solve some of the world's largest problems and challenges. In addition, everyone should be aware of the United Nations Millenium Goals (UNMG) and decide if our work can help address some of the major humanitarian challenges. The UNMG are listed at the end of this chapter.

Society needs to institute laws and ethical codes of conduct to guide the direction and impact of scientific and technological development. To ensure that engineering students and engineers become socially responsible, it is essential that they learn how to assess the impact of engineering designs and developments on people and on society, as well as the process of ethical decision-making. The curriculum of engineering schools should include these concepts and analytical approach. An over-arching principle to keep in mind is the dynamic nature of the issues to be included in such courses [Frize, M., 1996, 2003].

4.1 THE INTER-DEPENDENCE OF SCIENCE, TECHNOLOGY AND SOCIETY

There is a close relationship and interaction between these three concepts. Science can be defined as knowledge obtained by the systematic study of the structure and behavior of the natural world. Technology is the practical application of scientific knowledge. Society is defined as: (1) a group of people related to each other through persistent relations; (2) a large social grouping that shares the same geographical or virtual territory, subject to the same political authority and dominant cultural expectations [Wikipedia, 2011m].

Science and technology are not independent from each other. They are social processes that respond to a variety of economic, social, cultural, and political influences [Weiss, C., 2005]. Similarly, the diffusion of scientific knowledge and technological capabilities from a country to the rest of the world can give rise to a variety of policy and power issues; Weiss, C. [2005] argues that these can frequently become highly political issues. Examples can be seen in the questions of climate change, genetically modified foods, and the issue of foreign species invading new geographical areas, often taking over. The invasion of foreign plants and species can have a disastrous and lasting impact on local industries and create major stress to the sea, rivers, lakes, and land. Any scientific or technological change can have serious impact on society and on international relations. When examining such issues, it is important to consider the political, cultural, and economical aspects [Weiss, C., 2005].

4.1.1 EXAMPLES OF THE IMPACT OF SCIENCE ON TECHNOLOGY

Basic research on microwave radiation gave rise to the invention of the laser. The compact disc enabled the development of supermarket checkout counter technology. Advances in medical physics gave rise to the Cobalt 60 radiotherapy equipment, the Computed Tomography Scan (CT-Scan), and Magnetic Resonance Imaging (MRI).

4.1.2   EXAMPLES OF THE IMPACT OF TECHNOLOGY ON SCIENCE

The Hubble telescope allowed new measurements to be made of the universe and provided increased scientific understanding of our world. The same is true for the electronic microscope, which lead to several new scientific findings in the realm of cellular biology and in other scientific fields.

4.1.3   IMPACT OF SCIENCE AND TECHNOLOGY ON SOCIETY

Science and technology have a major impact on society. It is easy to see how developments like the cellular telephone and the internet dramatically changed how and when people communicate with each other. Major changes have been brought about in transportation, energy, the media and entertainment, medical devices, genetic engineering, stem cell research, and biotechnology. Many issues arising from the development of science and technology which are now global enterprises can affect all nations. Several discoveries have enabled to improve human health and life spans, economic status, and education; however, there is also increased pollution, environmental degradation, and a reduction of biodiversity. There is also a digital divide between developing and industrialized countries. This refers to access to internet, the development of websites and its content, ownership of digital technologies such as computers and cellular phones, although the latter is now becoming more prevalent in poorer countries, at least for their more affluent population. Another upcoming major issue is the nano-divide, referring to having access to the development of nanotechnologies. This is discussed in more detail later in this chapter.

4.1.4   EXAMPLES OF SOCIETAL RESPONSES TO GLOBAL ISSUES CREATED BY SCIENCE AND TECHNOLOGY

Climate change is expected to create rising seas as one of its consequences in future years. This brought about an alliance of small Island States to deal with the issue. Concerns about the environment have also inspired the negotiation of treaties such as the Kyoto and the Montreal Protocols. There are also World Trade Organization agreements. However, it is yet to be seen whether these agreements will be respected and implemented by all nations and lead to positive outcomes for the world. There are also agreements on intellectual property (IP) through the World Intellectual Property Office (WIPO). Again, not all nations adhere to IP rules and procedures. There are endangered species treaties, and discussions are occurring about AIDS and about the access to medication for poor nations who need it most.

Other issues which affect many nations are: Outsourcing and off-shoring, which refers to the employment of many people in poorer countries which takes jobs away from populations in industrialized nations. There is a proliferation of weapons mass of destruction, of nuclear arsenal, of biological warfare science. There is serious competition for arms, space, cloning, and reproductive technologies. Weiss argues that the capability to manage technology and innovation leads to economic and political power. In order to agree to share acquired knowledge, science, or technology, one nation can impose social, political, or cultural conditions on another [Weiss, C., 2005].

Factors which have a direct impact on development are: The opinion of the public on support for science and technology and whether tax dollars should be spent on research and development. Foreign policies regarding budgets, agendas, priorities, education, and training have an impact on what gets done. Relations between states impact the issue of migration workers, the recognition of professionals, the attendance at conferences, and the exchange of students and of experts between nations. The agreements to protect IP are also critical in the negotiations between nations on sharing knowledge.

Factors that have more of an indirect impact are: The marketplace, the economy, politics, and culture; laws and regulations which can restrict trade or enhance it; the mean income level of the population and its gross national product (GNP); availability of human resources in fields that are important to development; availability of jobs and the level of investment of private corporations in research and development.

Information technology and innovation create competition and a power struggle; an example is the competition between the Soviet Union, the USA, and Europe in space exploration. There are matters of security and sovereignty, and the threat of anarchy and terrorism. Other concerns are related to controlling the movement of people, of money, of information, of knowledge, of technology, and of diseases, pests, seeds, drugs, and nuclear materials. Other current global issues are computer viruses, hacking, and the invasion of privacy. There is also the question of access to wireless, internet, and cellular telephones in countries with autocratic regimes. It can be said that, even though signals can be scrambled, people find ways to communicate in spite of the strict control by some of the governments.

4.1.5 NEW EMERGING TECHNOLOGIES AND THEIR IMPACT ON SOCIETY

This section deals with technologies that have revolutionized the way in which information is developed, accessed, made available, stored, and transmitted. These include computers and software, communication devices, surveillance cameras, internet and the World Wide Web, the telecommunications industry, and satellites. Computers have many uses. They help run cars and appliances more efficiently; allow instantaneous access to massive amounts of information; allow one to pay bills and meet other banking needs, shopping online, do research, study, vote, and communicate. Computers and internet allow anyone to spread his/her ideas rapidly through the web. Computers allow businesses to record sales, track inventory, communicate with customers and suppliers, and store customer information. They provide many sources of entertainment and can be used to compose music, movies, videos, and games. They have revolutionized the way picture, sound, and language get published and communicated [Budinger and Budinger, 2006]. They allow physicians to visualize the patient, perform physical examinations, access laboratory work. Recent developments on clinical decision-support systems help in diagnosis, patient management, and monitoring of patients with a variety of medical conditions. A example is predicting preterm births [Frize and Yu, 2010] or helping parents make difficult decisions regarding the care of their infant in the neonatal intensive care [Weyand et al., 2011].

In spite of all the benefits enumerated above, several problems are also generated by technology: When computers are "down," the individual, the local economy, the country, and even the world can be affected. Because computers and software change so rapidly, information from a few years back may be inaccessible in newer computers. Major problems are related to security and to privacy, and we need to protect individual autonomy and privacy, especially on the Internet. Every click of a mouse to open a web site or to send emails, blogs, or postings on social media can be tracked, and the information sold to interested parties. There is an increased demand by the public that programs released by companies be free of security flaws, because break-ins to computers can have severe consequence on users. The use of cookies, which save user preference when visiting web sites, may also store the user's e-mail address and can be misused. Personal information can be stolen by hackers who break into universities, corporations, banks, military bases, and hospitals. As a result, one might be a victim of identity theft, credit card fraud, or bank account theft. Surveillance cameras appear in many public areas and in the workplace, as well as in homes, monitoring the perimeter of the home, the neighborhood, neighbors, visitors, and workers such as nannies or cleaning staff; they are also used to monitor children, pets, and the elderly. Although surveillance technology has some benefits, such as enabling older persons to live at home longer, it can also be pervasive and breach privacy rights.

Western governments have installed a major surveillance system that involves all individuals in these countries. The system called ECHELON is probably not well-known. ECHELON is a signals intelligence collection and analysis network operated on behalf of the five signatory states (UK, USA, Australia, Canada, and New Zealand). It has also been described as the only software system which controls the download and dissemination of the intercept of commercial satellite trunk communications. In a report, the European Parliament stated that ECHELON was capable of interception and content inspection of telephone calls, fax, e-mail, and other data traffic globally through the interception of communication bearers including satellite transmission, public switched telephone networks, and microwave links. Global positioning system (GPS) monitoring has benefits when a person is lost, but it allows secret tracking of individuals by anyone attaching a simple device to the system. This technology can locate an individual or a vehicle and its speed.

Personal medical records, now stored on computers, are open to hacking and privacy invasions, as they are often insecure and protected only through passwords. Telephones can be tapped. Small cameras and microphones can be hidden on a person to record another person's actions and words. Bridge toll plazas, stoplights, and other "smart" road devices can record license plate numbers. Although using surveillance cameras has law enforcement purposes, it goes against the basis of democracy and freedom. Monitoring in public areas such as airports is important, but it is more crucial to come up with preventive measures rather than simply catch wrongful acts. Employer and employees should develop principles and guidelines for computer monitoring, to protect the company while giving employees some freedom [Budinger and Budinger, 2006].

Another invasive technology is the new vehicle monitoring systems which can, among other things, log engine and driver performance, troubleshoot problems, and keep tabs on how and when the vehicle is being driven. For a rented car or a lease purchase, some systems installed in the car can send the information back to the owner of the vehicle.

The University of California published guidelines to help protect personal information: They suggest not to connect a computer having personal or sensitive information to the Internet; to establish a system of encryption and de-encryption for all originators and recipients of sensitive data; to de-identify all sensitive data stored on electronic files and store key codes in a file that cannot be accessed from the Internet [University of California, 2008].

4.1.6   OTHER PROBLEMS ARISE FROM INTERNET USE

Some examples are the following: harassment, stalking, pedophilia porn sites, hacking, hoaxes, scams, viruses, plagiarism, theft, deception, fraud, and spam. When regulations are introduced, individual rights can also be violated. Originally, hackers were trained to challenge the security of technology to pinpoint weaknesses; but eventually, they used their skill to cause problems of their own. Hoaxes on the Web can reach millions of people rapidly and burden the information system; this can decrease the speed of transmission of information, and are time-consuming to fix. Misleading information by journalists can be spread quickly by the internet. Internet users need to filter fact from fiction and assess the credibility of sites visited [Budinger and Budinger, 2006].

4.1.7   POLICIES AND REGULATIONS

There are government policies that encourage self-regulation of the internet industry and end-user voluntary use of filtering and blocking technologies in the US, UK, Canada and New Zealand. Criminal law penalties exist in some Australian states for internet web site providers who make content 'unsuitable for minors' available online. Some attempts have also been made in the US with various bills to ensure filtering software would be used in public internet access facilities such as libraries. More information on how the problem of child pornography is dealt with in several countries can be found at Wikipedia [2011g].

Government mandated blocking of access to content deemed unsuitable for adults were set-up in the Australian Commonwealth, China, Saudi Arabia, Singapore, United Arab Emirates, and Vietnam. Government prohibition of public access to the Internet still exists in countries like China, Cuba, Iran, and the United Arab Emirates. Protecting children so that they can use the internet safely and appropriately is an important issue, but this is difficult to control because children have access to many computers, not just those in their own home [Budinger and Budinger, 2006].

4.1.8   THE ENVIRONMENT

The Global Footprint Network, a 501c (3) nonprofit organization, was established in 2003 to enable a sustainable future, where all people have the opportunity to live satisfying lives within the means of one planet. This organization writes:

The Ecological Footprint is a measure of humanity's demand on nature. It measures how much land and water area a human population requires to produce the resource it consumes and to absorb its carbon dioxide emissions, using prevailing technology. Since the 1970s, humanity has been in ecological overshoot with annual demand on resources exceeding what Earth can regenerate each year. It now takes the Earth one year and six months to regenerate what we use in a year. Conceived in 1990 by Mathis Wackernagel and William Rees at the University of British Columbia, the Ecological Footprint is now in wide use by scientists, businesses, governments, agencies, individuals, and institutions working to monitor ecological resource use and advance sustainable development [Global Footprint Network, 2011].

4.1.9   SUSTAINABLE DEVELOPMENT

Because populations are growing, resources dwindling, and technologies growing, the capacity for mass destruction exists. Humans have a moral responsibility to care for the environment. An environmental resource which is publicly available or common like water is often abused and used indiscriminately. A severe shortage of water is predicted for the future, and currently one billion people do not have access to safe drinking water. Moreover, the water available in rich countries has been contaminated with antibiotics, phosphates, and many other pollutants which will have a detrimental effect on populations. Water is a life and death issue and engineers need to be part of the solutions to clean up the contaminated supplies, to help diminish the huge drainage of lakes for bottling water, for the manufacture of ethanol gasoline, and for other products.

It is important to maintain principles of sustainability in all undertakings. Operations that use natural resources must ensure these stay within the carrying capacity of the ecosystem, meaning that harvesting rates should not exceed the regeneration rate. Waste emissions should not exceed the assimilative capacity of the environment. The rate of exploitation of nonrenewable resources such as fossil fuels should be equal to or less than the rate of development of renewable substitutes such as solar, wind power or nuclear power. Each issue requires careful thought and planning.

4.1.10   POVERTY

Population statistics predict that by the year 2025, the population in western democracies is expected to be 8.6%, but in the rest of world it is expected to be 91.4%. Catalano, G. [2006, 5] cites a 1989 report by the United Nations Development Program [UNDP, 1989] claiming that, in 2006, "3 billion people have virtually no recourse to the basic necessities of life... "food, education, water, and sanitation." In a table on page 6, the author provides statistics obtained from the World Bank (2003; 2004), UNESCO (2003) and the United Nations (2003): 1.1 billion people lived on less than 1$ per day; 831 million were under-nourished; 104 million primary age children were not in school, of which 59 million (56.7%) were girls; 11 million children under 5 died each year; 1.197 billion people were without access to improved water sources and 2.742 billion were without adequate access to sanitation. It is interesting to contrast these dismal statistics with yearly expenses on luxury items in richer countries and observe that the latter, if spent on world poverty-related problems, could solve most of the world's most urgent ones; Catalano obtained the information from the Program of Action of the United Nations International Conference on Population and Development held in 2006 [UN, 2006]. Here is an inventory of expenses of luxury products for 2006: Expenses for makeup was 18 billion dollars, for pet food in Europe and US, 17 billion; for perfumes, 15 billion; for ocean cruises, 14 billion; and for ice cream in Europe, 11 billion dollars. This represents a total for luxury expenditures of 75 billion dollars a year. To meet the social or economic goals such as reproductive health care for all women would cost 12 billion dollars a year; to eliminate hunger and malnutrition, 19 billion; for universal literacy, 5 billion; for clean water for all, 10 billion; and to immunize each child, 1.3 billion. The expenses to solve the world's hunger, health, and access to education would be 47.3 billion dollars. This would be achieved by eliminating makeup, perfumes, and ice cream from luxury expenses. Engineers could choose to provide their expertise to organizations such as engineers without borders (EWB), or to the United Nations Development Program (UNDP), instead of lending their time and effort for the production of the luxury items mentioned above.

4.1.11 FAST EMERGING NANOTECHNOLOGIES, ETHICS, AND SOCIETY

Nanotechnologies are developing at an incredible rate, in fact much faster than the studies that could provide insight into their potential harm. It is known that nanotechnologies can provide dramatic advances in health care through drug delivery systems, bone repair, diagnostic tools, and therapies for cancer, diabetes, and other chronic diseases. The general impression is that their development is little cause for concern, both regarding human health and the environment. However, it is a fact that the particle form of these technologies can create health and safety problems. Mills and Fleddermann [2005] state that if we are to benefit most from nanotechnology, we need to pay close attention to its societal and ethical implications. The authors mention that humans have used nano-particles for many centuries; examples given are: glazes on Ming ceramics, Etruscan eyeliner, and stained glass windows. Modern nanoscience and nanotechnology have arisen following inventions such as the scanning tunneling and the atomic force microscopes, leading to the capability of manipulation and fabrication.

In traditional materials, the surface particles are highly reactive, but the bulk of the material is relatively inert and surface atoms are a small fraction of the total number of atoms in the material. However, in nanomaterials, a 30nm particle has 5% of atoms on its surface, but a 3nm particle has 50% of its atoms at the surface. So the bulk form of a material can be safe, but its nanoform can be toxic. For example, carbon nanotubes may be carcinogenic, whereas the diamond form of carbon is inert. It is known that colloidal quantum dots can penetrate living cells, so if used for therapeutic intent, they could possibly circumvent the body's defense mechanism. Nanomaterials can also pose environmental hazards as non-biodegradable pollutants, and they can transport contaminants.

Also there are privacy issues; when genetic discrimination becomes a possibility, nanochannels will be able to sequence DNA in minutes. There is also the matter of micro-surveillance, as these tiny materials can be hidden in the smallest holder. Another question is, will there be a nano divide like the digital divide, where poor countries have little access to computers and internet technologies? Will there be twenty-four hour access to information through a wrist computer? Will nanobots be used to make diagnosis or therapy in focused areas inside the body? Will there be an increased longevity because of access to nano-medicine? This can lead to nano-haves and nano-have nots. Will nanotechnologies have an effect on traditional markets as a disruptive technology? How will foreign cultural values consider this technology? There is a serious lack of ethical framework to ensure safe implementation of these materials and of the technologies that use them. There is a need for regulation for these types of particles, yet as we know the legal frequently lags technological innovation. There is also a major challenge in protecting intellectual property in a nano-age. In June 2005, there were 7696 patents issued with the word nano in their description. The investment from several countries is huge. Mills and Fleddermann mention that in 2005, there were 55 companies manufacturing carbon nano-tubes and that one company (Mitsui Corporation) built a plant that began large production (120 metric tons) in 2004. At that time, a few universities (Rice and Rochester) began safety studies, examining the toxicology of several nanomaterials [Mills and Fleddermann, 2005].

Education will be a key to success in controlling this development and its impact on society. Who will need to be educated about this? All stakeholders will need to be aware of nanotechnologies and this includes judges, lawyers, journalists, the public, and children in school. Engineers, scientists, social scientists, ethicists will also need this knowledge. If regulation and safety are not controlled, it could very well happen that nanotechnologies will follow the same controversial path that genetically modified foods did, with major resistance developing in Europe, Canada, and elsewhere.

4.1.12 RESEARCH AND IMPACT ON SOCIETY

Holbrook, J. [2005] mentions that the National Science Foundation, the major funding agency for research in science and engineering in the USA, has recently added new criteria for selecting successful grant proposals for funding. In addition to intellectual merit of the proposal, researchers have to describe the broader impact of the proposed activity, in relation to the impact on education, infrastructure, diversity, and societal benefits. This has created a tension between scientific autonomy of the researcher and societal control of the direction and scope of a scientific activity. Some of the questions around this issue are: Can science stand on its own without any concern for society and people? Are there not times when scientific knowledge distorts lived reality? Is technological power not only a boon, but also a danger to human welfare? Vannevar Bush, the opponent of this new criteria argues that the concept of basic research comprises a lack of concern with practical ends to the work, as these ends are normally the province of applied research. Holbrook, J. [2005] states that Daniel Sarewitz supports scientific research that serves the public interest; he quotes this author: "The general idea is to graft mechanisms onto the system that creates a stronger motivation for pursuing, and better tools for recognizing and measuring, direct contributions of science to societal goals" [Sarewitz, D., 1996]. Holbrook, J. [2005] also mentions in his article that "Gingrich, Sarewitz, Pielke, Byerly, and Stokes all agree that in order to justify continued public investment in scientific research, the scientific community must adopt a new model of scientific inquiry that incorporates intellectual considerations of the nature of scientific research with considerations of societal benefit" [Holbrook, J., 2005].

The United Nations Millenium Goals: The United Nations Secretary-General, Kofi A. Annan stated:

> We will have time to reach the Millennium Development Goals – worldwide and in most, or even all, individual countries – but only if we break with business as usual. We cannot win overnight. Success will require sustained action across the entire decade between now and the deadline. It takes time to train the teachers, nurses and engineers; to build the roads, schools and hospitals; to grow the small and large businesses to create the jobs and income needed. So we must start now. And we must more than double global development assistance over the next few years. Nothing less will help to achieve the Goals" [UN, 2006].

Engineering student and engineers should be aware of these goals and do their best to include them whenever possible in their work, especially in the developing world (UN). The goals are:

1. Eradicate extreme poverty and hunger: Reduce by half the proportion of people living on less than a dollar a day. Reduce by half the proportion of people who suffer from hunger.

2. Achieve universal primary education: Ensure that all boys and girls complete a full course of primary schooling.

3. Promote gender equality and empower women: Eliminate gender disparity in primary and secondary education preferably by 2005, and at all levels by 2015.

4. Reduce child mortality: Reduce by two thirds the mortality rate among children under five.

5. Improve maternal health: Reduce by three quarters the maternal mortality ratio.

6. Combat HIV/Aids, malaria and other diseases: Halt and begin to reverse the spread of HIV/AIDS. Halt and begin to reverse the incidence of malaria and other major diseases.

7. Ensure environmental sustainability: Integrate the principles of sustainable development into country policies and programs; reverse loss of environmental resources. Reduce by half the proportion of people without sustainable access to safe drinking water. Achieve significant improvement in lives of at least 100 million slum dwellers, by 2020.

8. Develop a global partnership for development: Develop further an open trading and financial system that is rule-based, predictable and non-discriminatory, includes a commitment to good governance, development and poverty reduction—nationally and internationally. Address the least developed countries' special needs. This includes tariff- and quota-free access for their exports; enhanced debt relief for heavily indebted poor countries; cancellation of official bilateral

debt; and more generous official development assistance for countries committed to poverty reduction. Address the special needs of landlocked and Small Island developing States. Deal comprehensively with developing countries' debt problems through national and international measures to make debt sustainable in the long term. In cooperation with the developing countries, develop decent and productive work for youth. In cooperation with pharmaceutical companies, provide access to affordable essential drugs in developing countries. In cooperation with the private sector, make available the benefits of new technologies—especially information and communications technologies [United Nations, 2011].

4.1.13   EXAMPLES OF ISSUES OF CONCERN IN DEVELOPING COUNTRIES

Although several of the issues discussed for industrialized countries also apply to developing countries, there are specific concerns to consider for the latter. Some companies test some of their new products in developing countries because they are not restricted by rules and ethical guidelines that would not allow these tests in a western country. A new product or procedure, to be tested on humans, must be compared to the best existing product or procedure in order to be accepted by ethics review boards. Take for example the case of the testing of a new surfactant (Surfaxin) in Bolivia. Respiratory distress syndrome arises because of the lack of surfactant in the lung of a premature infant, which is a complex compound made of protein and fat which help the air sacs inflate with air. The current treatment for this condition is to administer a surfactant drug into the lungs. In 2001, the US-based Corporation *Discovery Labs* sought approval of the *Food and Drugs Administration* (FDA) for a study of Surfaxin. They conducted a placebo-controlled trial in Bolivia, which means that some of the infants received the surfactant, while others were given a placebo. This would be strictly prohibited in the USA and in other western countries. The company's reasoning was that infants were not receiving this treatment before the study, and now some infants would have access to the treatment. This is unethical according to the codes of ethics and ethical theories [Flaherty and Struck, 2000].

Another example was the testing of a reduced amount of AZT (Azidothymidine), the first HIV antiretroviral drug produced in 1987, against a placebo. In 1994, a study in France and in the US had showed that AZT reduced HIV vertical transmission from mother-child by a factor of two-thirds. The new trial planned to test the effectiveness of short term, modified dose of AZT for developing countries, using a control group receiving a placebo, and the experimental group receiving the modified AZT. In 1997, there were 18 randomized clinical trials (NIH and CDC funded) which tested 17000 HIV-positive women in Asian and African countries [Wolfe, L., 1997].

In Nigeria, there was an epidemic of meningococcal meningitis in 1996. Pfizer performed foreign medical research of the oral antibiotic drug trovafloxacin "Trovan" at the Aminu Kano Teaching Hospital in Nigeria. Two groups of 100 children and infants with meningitis were used in the trials. The first group was given oral formulation of *Trovan,* while the control group was administered a low dosage of an antibiotic (ceftriaxone, commonly called *Rocephin*), made by one of Pfizer's competitors, and approved by the FDA for the treatment of meningitis. Five of the children given trovafloxacin died after the treatment while others developed mental and physical problems such as arthritis and hearing loss. Six children died while taking the comparison drug. In 2000, the *Washington Post* launched a series of stories describing clinical trials in developing countries done by US researchers. US researchers were "not ensuring that overseas study subjects gave informed consent" [Stephens, J., 2006]. Other effective and approved alternative drugs were available at the time of these trials, but the families were not given a choice, not even to refuse the drug. At the time of the experiment, *Trovan* was at the experimental stage and was not approved by the FDA; it had not yet been tested on humans. There was falsification of a document to have the drug approved, to put it on the market, and to make a profit out of the current epidemic or from another potential one. It is important to note that only 68% of the population of Nigeria is literate [CIA, 2011]. Such studies in developing countries should be avoided due to the high level of illiteracy and an uneducated population. For more details on this study, see: Ready, T. [2001, 2000], Wise, J. [2001], and Stephens, J. [2006]. A US Bill called "Safe Overseas Human Testing Act" was introduced in the US Congress in 2005-2006 to establish guidelines for trials in developing countries, but unfortunately it never became law [govtrack.us, 2006]. It is important to remember the statement in paragraph 29 of the Helsinki Declaration:

> *"The benefits, risks, burdens and effectiveness of a new method should be tested against those of the best current prophylactic, diagnostic, and therapeutic methods. This does not exclude the use of placebo, or no treatment, in studies where no proven prophylactic, diagnostic or therapeutic method exists* [WMA, 2004].

The World Health Organization also made a statement regarding this issue: *"The ethical standards applied should be no less exacting than they would be in the case of research carried out in [the sponsoring] country"* [CIOMS, 1993].

In conclusion, developing countries need to make strict rules and laws to protect their populations from being exploited by organizations, researchers, and corporations from their own country or from other countries. The fact that no treatment exists for certain conditions does not warrant using a placebo in lieu of an existing effective treatment. More importantly, informed consent must be obtained and experiments must respect all the ethical guidelines in place in developed countries.

CHAPTER 5

Gender, Culture, and Ethics

5.1 GENDER AND ETHICS

Another important aspect of ethical behavior in engineering is how persons from under-represented groups are treated by employers, peers, or employees. There must be zero tolerance for harassment or discrimination in the workplace and in engineering schools. Raising awareness of what harassment and discrimination are and how to eradicate them is an important step in creating an equitable and safe environment for women and for other under-represented groups in the workplace [Frize, M., 1995]. The Association of Professional Engineers of Ontario (PEO), which came into existence in 1922, made these actions misconducts in 1990 for which engineers can be disciplined. The document can be found on the PEO website under Publications, Professional Engineers Act (R.R.O. 1990, Reg. 941, s. 72 (1); O. Reg. 657/00, s. 1 (1) [PEO , 2011].

Another aspect of justice is to ensure equitable access to medical services, which includes not only both sexes, but also populations from under-represented groups. Several health issues differ by sex and by racial background. Issues are also dependent on age. For example, concerns for teenage women could be: eating disorders, self-image, date rape, and pregnancy. Later on, issues are related to family planning, balancing family and career, being the sandwich generation looking after aging parents and children. These situations can lead to stress and burn-out. Some of these factors can also affect men, but in general it is still women who have the overwhelming burden of these life responsibilities. Physical abuse is still a problem affecting some women during their childhood and adult life. After menopause, women are twice as likely as men to develop osteoporosis; many more women than men suffer from depression; they can also develop breast cancer, and face difficult decisions about hormone therapy and hysterectomy. Men can face erectile dysfunction and prostate cancer. Both sexes face heart disease and diabetes. In view of the differences, it is critical that research, especially projects using public funds, be spent equitably on diseases that affect women and men. There is also a difference in access to medical services, not only for rural or remote populations compared to city dwellers, and for the poor compared to the rich, but also for women and children compared to what men receive. Disparities are also enormous between developed and developing nations. For example, comparing the life expectancy and health care expenditures per person per year between Canada and Tanzania, one can see a huge disparity: In the mid 2000s, in Canada, the life expectancy for women was 83 years and for men 78; Canada spent $2989 (Can.) per capita in 2006 on health care. In Tanzania, life expectancy in a similar year was 47 years for men and 49 for women; the yearly expenditure per capita on health care was 4US$. Other factors, in addition to life expectancy, that are greatly affected by the wealth or poverty of a country are: child mortality, death of mothers during childbirth, access to education and literacy, clean water, and a safe environment. Strategies need to focus on major emerging issues; for example, obesity in western countries, HIV AIDS, malaria, and resistant tuberculosis in African countries. Community outreach and education should be a priority for intervention programs, and it is critical to raise the awareness of physicians and all health care workers, administrators, politicians, policy makers, funding agencies, engineers, and researchers for change to occur. Over medicalization of natural reproductive processes is also a matter of ethics. Mentoring and supporting activities can help bring positive changes to the health care system in matters of justice and equity in access and development.

The media reported several issues related to women's health in the 1990s, some of which created serious health complications. An example is the problems that occurred with breast implants. There was an article mentioning that women lose out on care after a heart attack. One study showed that men get more aggressive treatment and more of them improve after a heart attack, while more women than men decline towards congestive heart failure and women also had more chest pain. Although women had more doctor visits, they had less diagnostic tests prescribed. The Minister of health of Ontario at the time (E. Witmer) said that male bias in health care kills women (Globe and Mail, July 28, 1997).

Women experience inconsistency in accessing care, treatment, and drug therapy. The system was principally designed by men, for men. Sue Rosser said: "In our culture, the institutionalized power, authority and dominance of men frequently result in acceptance of the male world view or androcentrism as the norm" [Rosser, S., 1989]. It is only with time and a strong will, confidence and position that people can stray from the traditional model and use new qualities and approaches, especially in non-traditional fields.

Another consideration is the way in which medical terms are used. Here are some examples of sexist and pejorative terms still used in some medical texts: Multiple miscarriages is sometimes called 'habitual aborter;' a miscarriage is a 'spontaneous abortion' or 'incompetent cervix,' or 'blighted ovum.' It is interesting that such terms are not applied to men's sperm lacking motility or for premature ejaculation. Regarding the difficulty in conceiving, terms like 'hostile cervical mucus' or 'inadequate luteal phase' can be found; the uterus has been referred to as a 'hostile environment' for the early embryo, ovaries as 'senile,' 'barren,' or 'inadequate' [Callahan, J., 1995].

Regarding research, it is important to develop an equitable process on the following matters: Who decides what research questions should be asked; for example, some effort is needed to improve the detection and treatment of both breast cancer and of prostate cancer. What information should be collected, who interprets it, who sets the priorities, what and who gets funded, what gets published? How can we balance what society needs, what researchers are interested in, and what funding agencies tag as priorities?

Concerning access to medical care, there are other examples of inequities. For example, in the Province of Alberta, on the matter of hip replacement, an Ottawa Citizen article on March 16, 2003 talked about a two-tier system: One list of patients was done in three weeks; this was a private service, for a fee. On a second list, patients with a public access to health care would wait typically one year.

However, on the second list, there was a short list of patients from the Workman's Compensation Board, prisoners, Royal Canadian Mounted Police personnel and military personnel. The long list consisted of all other Canadians who needed this surgery in that Province.

When research projects are planned, researchers must ensure inclusiveness of both sexes and representing various racial groups in the community so that the benefits can apply to a broad section of the population. Until recently, most clinical studies were performed exclusively on male subjects. Some of the new studies seem to be more inclusive of both sexes and of various racial backgrounds. However, gender inequalities still exist in all parts of the world, developing and developed, to various degrees. This pertains not only to research but to access to clean water, food, health care, education, and all aspects of life and security [Moreno, J., 2005]. Consequently, in planning new studies, researchers and activists must assess how the planned work or project will affect all people, women, men, and children, and how the work can improve life on earth for everyone.

5.2   CULTURE AND ETHICS

Two examples are provided in this chapter. One concerns cultural attitudes in Japan, the other is a case study regarding an Aboriginal child in Canada. Both examples illustrate how attitudes affect how health care services are perceived by patients and their family, and the end result for patients.

Masahiro Morioka, in his article *Bioethics, and Japanese Culture: Brain Death, Patients' Rights, and Cultural Factors* [Morioka, M., 1995] presented the history of heart transplant in Japan. The first heart transplant in Japan was done in 1968, a year where such surgeries were being done in the USA and in South Africa. The patient in Japan lived 83 days. The death of the patient was followed by an accusation by a citizen group of illegal human experimentation and of exercising dubious judgment with respect to a donor's brain death. The consequence of the protest was that heart transplantation became a taboo for the next fifteen years in Japan. One major barrier to this type of medical procedure was that, in Japan, the concept that brain death means human death was rejected; people also expect a dead person to be whole in order to pass into the next life; therefore, any type of major surgery or organ removal is deemed unacceptable. The influence of traditional Japanese culture also created a clash with the autopsy system which was connected with transplantation procedures.

Another challenge arising at that time was that the concept of a patient's rights and consent was difficult to apply in a country where paternalistic medicine was still practiced; physicians did not accept being questioned by patients on their orders or diagnosis, and some would not even provide the truth about terminal conditions. If you asked a doctor about the medication prescribed, you could get scolded and sometimes told to be silent. Some doctors refused to give patients important medical information concerning their illness. Many physicians think that patients should live in the web of warm-hearted consideration shown by the surrounding people, and that it is a good custom that a patient leaves his/her decision making to intimate others. In his article, Morioka, M. [1995] states that "recent studies by Naoko Miyaji, an anthropologist in Japan, show that the most frequent responses from the physicians she interviewed were: The physicians themselves wanted to know the truth, but in the case of their patients, they do not tell the truth. Miyaji surmises this as one of the reasons for the low rate of truth telling in Japan (Miyaji, N. [1994], from Morioka, M. [1995]). The 1992 survey showed that only twenty percent of patients with terminal cancer were told the truth; doctors lied to eighty percent of the patients with this condition. Patients could not find important information and could not say anything against the doctor (Miyaji, N. [1994], from Morioka, M. [1995]). This led to a distrust of physicians by the population and rendered research more problematic to carry-out. At the time, most of the people on the eighty ethics committees in Japan were professors of the medical schools. This also gave rise to mistrust and, consequently, to fear that transplants would be done with minimal information provided to the family of the donor. People feared that they would be pressured by doctors to sign permission for the organ retrieval, and that perhaps they would even be threatened if they did not agree to sign.

The cultural factor regarding transplants is quite opposite in North America. Americans and Canadians think of organs as replaceable parts, in keeping with the mind-body dualism philosophy. In Japan, every part of the body is a fragment of the deceased mind and spirit and thus the body must be perfect. If parts are missing, the soul becomes unhappy in the next world and this sadness can affect the family members. This vision is also shared by other Asian cultures, specifically adherents of Confucianism and of Shamanism. More recently, in that region, there have been activists fighting for patients' rights, for informed consent, and for insisting on being told all one needs to know prior to a procedure or to a research experiment. Self-determination has also become an important concept in Japan. For their part, physicians began to adopt a more western style of interactions with patients. Organ transplants are getting more support; however, the concept of brain death remains an issue [Morioka, M., 1995].

5.3   CRITICAL ILLNESS DECISION FOR AN ABORIGINAL CHILD IN CANADA

The second example illustrates a situation of cultural pluralism. In their article *Perspectives on health and cultural pluralism: ethics in medical education,* Harold Coward and Gwen Hartrick, at the University of Victoria (2000, 261-265) present the Case of K'aila, an Aboriginal child born at home in Alberta. At three months old, the pediatrician informed the parents that K'aila would die if he did not get a liver transplant. But the outcome of the surgery was uncertain as to whether the child would live or die after the procedure. This case illustrates ethical, cultural, and religious issues and reinforces the need for interdisciplinary approaches in medicine that includes views from the humanities, the social sciences, and ethicists. Medicine, as a culture, comprises a common body of knowledge, history, value system, rules, and norms. Medical specialists have special rights and privileges, including the power to determine the practice and regulation of bestowing its special knowledge, which is a form of power in our Society. A very difficult situation occurs in medical ethics when the parents refuse a life-saving treatment for a child. When such cases turn into a lawsuit, saving a child's life is usually supported against the parents' religious or cultural beliefs, based on the principle that children have a right to mature into autonomous persons who can choose their own beliefs and values.

K'aila had jaundice at birth which was treated, but he developed more serious liver problems when just a few months old. The pediatrician's viewpoint was that the child would die unless a liver transplant was done and cited a potential survival rate of 80 to 85%. The authors of the article explain the parents' viewpoint as follows:

K'aila's parents were well educated. As time passed they felt pressure to approve a liver transplant for K'aila. However, with the help of others, they studied about transplantation and the related problems of rejection and susceptibility to infection. In their view, the five-year survival rate after transplantation seemed close to 60% or 65%. Moreover, in their cultural and spiritual beliefs, they would be committing a serious error to recreate their son's body using an organ from another person. In their Aboriginal beliefs, this would bring the spirit of the other person into their child, which was not acceptable to them. Their awareness of a lower survival rate, together with their Aboriginal cultural beliefs, made them conclude that they would refuse a transplant for their son. A court application was made requesting that K'aila be taken into custody so that the transplant could take place without the parents' consent. In the end, the court rejected the application for custody and upheld K'aila's parents' right to make their own decision, not because of their spiritual beliefs, but because of the uncertainty over the side effects and ultimate outcome of the liver transplant. The parents returned home and K'aila died when he was eleven months old.

This case demonstrates that physicians need to learn, while in medical school, or afterwards if this was not part of the curriculum, to be sensitive to other cultural beliefs and traditions. Of course, if the outcome in this case had been more certain, in a full recovery of the child, the court would have likely judged in favor of the pediatrician and a transplant would have been done. But in this case, when the clinical outcome was uncertain, and the pain, suffering, and complications resulting from the surgery were real, the physician should have had a more sensitive approach and supported the parents in their decision. Medical schools should discuss this type of case and encourage physicians to be more sympathetic in considering cultural values that differ from their own. This does not mean that they have to agree with all aspects of other cultures, but they should at least be aware that they exist and try to be more sensitive and understanding. Physicians also need to take into consideration quality of life issues when making recommendations on treatment options for their patients.

CHAPTER 6

Data Collection and Analysis

This chapter discusses issues to consider when measuring living systems, that is, humans or animals, and the principles defining a good scientific approach. Definitions are provided on what is a Law, a Hypothesis, a Model, and a Theory. Challenges regarding data collection and analysis are presented. Finally, the importance of avoiding potential bias is discussed.

6.1   MAJOR ISSUES IN MEASURING LIVING SYSTEMS (HUMAN EXPERIMENTATION)

Some measurements of living systems are difficult to achieve because of the frequent **inaccessibility** of the variables to be measured. An example is the measurement of blood flow; if we were to measure this directly, we would be using a flowmeter around a blood vessel. Of course this can be done during surgery when the vessels are exposed. But in general, indirect methods must be devised to approximate the measurement.

Another issue is the **variability** of the data, not only from one patient to another, but also for the same patient at different periods of time. Other factors of variability are gender, race, age, size, weight, and various physiological or pathological conditions. Lack of knowledge about inter-relationships and the inter-action among physiological systems are also major challenges. Biological systems are very complex and mostly non-linear. It is also important to identify **confounding variables**, as they can bias the analysis. There are unknown factors that can affect the measurements or the analysis. Incomplete data sets and **missing values** can be a major obstacle when using machine learning and statistical approaches for the analysis of the data, and it is often necessary to either eliminate cases with missing values from the data set, or to find a reliable manner to replace them. One more difficulty concerns the potential effect of transducers or sensors on the measurement; there can be a disturbance of the measurand (the quantity being measured), artifacts, and bias introduced in the data collected. An example is when using thermocouples or thermistors to measure the temperature of the body, as they are in contact with the part being measured. One way to avoid this problem is to use an infrared camera to measure temperature, as it does not touch the body.

6.2   PRINCIPLES TO CONSIDER WHEN PERFORMING STUDIES WITH HUMAN SUBJECTS

Hospitals and universities have their own application form and ethics review board or committee, as well as their own process to approve or not an application for a new study. Studies must conform to generally accepted scientific principles and be based on solid previous evidence established either by computer or by animal studies. Sometimes it is not possible to carry-out such previous work, and this can be justified or explained in the background part of an application for a new study. Studies must keep in mind at all times the safety of the subjects; in some experiments, a physician or nurse may have to be present. The importance of the objective of the study must be considered and compared to the inherent risk to the subjects to be recruited. As seen in Chapter 2, the subjects must be thoroughly informed about what is to be done and the potential harm or risks before they are asked to sign the consent form. The confidentiality and privacy of the subjects must be preserved at all times.

6.3   CONSIDERATIONS PRIOR TO DECIDING ON THE STUDY

Consider the purpose of the experiment, the benefits expected, and whether the study is for diagnostic or therapeutic purposes, and whether it is elective or essential. Are the consequences known? What are the probable and possible risks? Are there alternative treatments or therapies? Would there be consequences in refusing to participate?

6.4   USING A GOOD RESEARCH METHODOLOGY

A good scientific method "begins with the foundations of rational thought processes. These include a clear conceptualization, definition, and inference. Concepts, clearly defined, are used in the inference process for building propositions, hypotheses, models, theories, principles, laws…(these terms overlap somewhat and mean different things to different people)" [Emory, W., 1980]. The process should be to build, then test until satisfied with the results. This must be followed by a careful analysis of the data or of the results, and a proper conclusion should be drawn. Generalizations or broader inference than is warranted by the results must be avoided. That is sometimes referred to as reductionism.

The methodology must be clearly described so that someone else can repeat the study. The steps must be planned in a way to get as objective results as possible, and the sample should be representative of the population to be studied. Personal or political bias must be avoided. If flaws in the methodology design are found, then this should be mentioned in the reporting of results and an explanation must be included on the potential impact on the findings. Methods of data analysis should be carefully selected to produce significant results. Most importantly, the reliability and validity of the data must be checked. When reporting the findings and the conclusions, this must be accompanied by the degree of confidence. Conclusions should be applied to the data and not broadened or extended to a wider population, especially when they are based on a small sample. The integrity, reputation, and experience of researcher are assets in any experiment. A definition of bias follows:

"A bias is a prejudice in a general or specific sense, usually in the sense for having a predilection to one particular point of view or ideology... A bias could, for example, lead one to accept or not-accept the truth of a claim, not because of the strength of the claim itself, but because it does or does not correspond to one's own preconceived ideas [Wikipedia, 2011a]."

6.5   DEFINITIONS THAT HELP DEFINE HOW THE PROBLEM IS STATED

A **concept** is defined as "an abstraction of meanings from reality to which we assign some word or words in order to be able to communicate about it" [Emory, W., 1980]. Kerlinger states that "a concept expresses an abstraction formed by generalization from particulars" (cited in Emory, W. [1980]). Holt et al. argue that "concepts are terms that refer to the characteristics of events, situations, groups and individuals that we are studying in the social sciences" (also in Emory, W. [1980]).

Hypotheses statements are made using concepts. We devise measurement concepts by which to test a hypothesis statement; for example, the concept of distance can be used to measure attitudes to something; or use a threshold to separate results, or a concept in perception.

Constructs are frequently used in the social sciences or in qualitative type research; they are an image or an idea applied to a specific research or to build a theory. Constructs are more complex than concepts as they are usually a combination of simpler concepts. The idea or image is often not directly subject to observation; examples are in the measurement of intelligence or of motivation.

"A hypothesis is a proposed explanation for a phenomenon. A scientific hypothesis must be able to be tested and be based on observations (or measurements) or are extensions of scientific theories" [Wikipedia, 2011e]. A descriptive hypothesis is a proposition that typically states the existence, size, form, or distribution of a variable. An example is: American cities are experiencing budget difficulties; cities represent the case, and budget difficulty is the variable. Another example is the statement that 90% of babies in the Neonatal Intensive Care Unit survive; babies represent the case and survival is the variable. A relational hypothesis is a statement which describes a relationship between two variables with respect to the same case. For example, imputing missing values using a hybrid k-nearest network and an artificial neural network will have a performance equal to or better than the mean or random imputation methods [Ennett et al., 2004].

There are three conditions to be respected to create a good hypothesis: (i) Be adequate for its purpose; it clearly states the condition, size, distribution, or expected result; (ii) be testable using available techniques; and (iii) it must be better than its predecessors and require fewer conditions or assumptions [Emory, W., 1980].

A **research question** is a statement of the problem to be resolved; for example, can artificial neural networks be used to predict pre-term birth with a better performance than existing methods? Or can we predict repeat injury in a population database of childhood injuries and identify risk factors that could lead to a repeat injury prevention program?

To differentiate a **theory** from a hypothesis, the former tends to be abstract and involve multiple variables, while a hypothesis tends to be simpler, such as a two-variable proposition involving concrete instances. Example of a theory is the theory of elasticity. A theory is valid for all times or until proven to be incorrect.

Laws: "A physical law, scientific law, or a law of nature is a scientific generalization based on empirical observations of physical behavior. They are typically conclusions based on the confirmation of hypotheses through repeated scientific experiments over many years and which have become accepted universally within the scientific community. However, there are no strict guidelines as to how or when a scientific hypothesis becomes a scientific law" [Wikipedia, 2011k].

For example, Boyle's Law: $V_2/V_1 = P_1 / P_2$, if T is same; or Charle's Law: $V_2 / V_1 = T_2 / T_1$ were established by experiment and will be valid for all time.

A **model** is often seen as an all-purpose word for explaining relationships between or among concepts. So we need a conceptual framework to describe a model. Sources to develop a model can be theories, laws, hypotheses, or principles. Models are created by speculation about processes that could have produced observed facts. For example, we can develop a model of injury prevention by having studied the most common factors that lead to injuries. This would be based on identifying the variables that lead to the injury type under investigation. The important variables found by the analysis become the basis for development of the prevention model.

Humans construct theories in order to explain or predict phenomena. "In many instances, this is seen to be the construction of models of reality. A theory makes generalizations about observations and consists of an interrelated, coherent set of ideas and models" [Psychology Wiki, 2011].

6.6   DATA COLLECTION

This is a critical step and one of the most problematic in many cases. Access to medical or clinical data is not a trivial matter. Moreover, data may have important missing values or variables. Audits must be done to check the quality and integrity of the data, to identify outliers and check whether these can be deleted or not. Acquiring data can be very expensive, especially if it is done with patients in a clinical context. Current privacy laws also make it considerably more difficult to gain access to patient data, even if identifiers are removed prior to accessing the data.

6.7   ANALYSIS AND INTERPRETATION

Care is needed when interpreting and presenting the results. Interpretation of results is an important step, but must only be provided if there is sufficient evidence for the inference (deduction) being made. Bias or reductionism must be avoided. Reductionism refers to the action of reducing the nature of complex things to simpler or more fundamental thing. This can be said of objects, phenomena, explanations, theories, and meanings [Wikipedia, 2011l]. Finally, a good knowledge of statistical analysis approaches is very helpful for clinical data analysis.

Conclusion

Engineers must keep abreast of changes; new technologies crop up faster than our codes, laws, and political systems to deal with change. Technological development brings major benefits and hazards in its wake. Each of us can do our part to ensure a positive impact for society and that our efforts help to solve some of the world's largest problems and challenges. Society must institute laws and ethical codes of conduct to guide the direction and impacts of the developments. But where such guides do not exist, say for emerging issues, it is important to apply ethical theories and codes of ethics to assess how to deal with new moral dilemmas. Instructors of ethics courses must ensure that engineering students become socially responsible engineers and they must learn how to assess the impact of their work on people and society. There should be a guideline on how to include these important concepts in our engineering curriculum. An over-arching principle to keep in mind is the dynamic nature of the issues to be included in such courses.

Bibliography

Andrews, G.C. (2005) *Canadian Professional Engineering and Geoscience: Practice and Ethics*, 127–128, 130, 126–127, 128–130, 141–144. Toronto, Thomson Nelson. Cited on page(s) 6, 7, 8, 9

ASME (2011) `http://sections.asme.org/colorado/ethics.html/` (accessed October 18, 2011). Cited on page(s) 10

Badylak, S.F. and R.M. Nerem (2010) "Progress in tissue engineering and regenerative medicine." *PNAS*, vol. 107(8), 3285–3286. Also available at: `www.pnas.org/cgi/doi/10.1073/pnas.1000256107` (last visited October 2011). Cited on page(s) 34

Beauchamp, T.L. and J.F. Childress (2009) *Principles of Biomedical Ethics*. Oxford University Press. Cited on page(s) 22

Biojewellery (2011) `http://www.biojewellery.com/` (accessed August 01, 2011). Cited on page(s) 32

BMES (2011) `http://www.bmes.org/aws/BMES/pt/sp/ethics/` (accessed October 18, 2011). Cited on page(s) 10

Buchanan, A. (2008) "Enhancement and the Ethics of Development." *Kennedy Institute of Ethics Journal*, Volume 18, Number 1, March, 1–34. Published by The Johns Hopkins University Press. DOI: 10.1353/ken.0.0003 Cited on page(s) 30

Budinger, T.F. and M.D. Budinger. (2006) *Ethis of Emerging Technologies: Scientific Facts and Moral Challenges*. Hoboken, John Wiley & Sons, 59–60, 291–292, 294, 406, 418, 419, 81–85, 90, 96. Cited on page(s) 11, 24, 30, 33, 38, 39, 40

Bugliarello, G. (2010) "Ethics of medicine, biology and bioengineering at the new critical crossroads for our species–beyond Aristotle and Hippocrates." *Ethics in Biology, Engineering & Medicine-An International Journal*, Vol. 1(1), 3–8. DOI: 10.1615/EthicsBiologyEngMed.v1.i1.20 Cited on page(s) 34

Callahan, J. (1995) *Reproduction, Ethics, and the Law: Feminist Perspectives*. Indiana University Press. Cited on page(s) 48

Canguilhem, G. (1989) *The Normal and the Pathological*. New York, Zone Books. Cited on page(s) 31

Catalano, G. (2006) *Engineering Ethics: Peace, Justice, and the Earth.* Morgan & Claypool Publishers. (`www.morganclaypool.com`). Cited on page(s) 41

The Parliamentary Centre (Parlcent) (2000) *Controlling Corruption, A Parliamentarian's Handbook* Second Edition, September 2000. `http://www.parlcent.ca/publications/pdf/corruption.pdf` (accessed August 01, 2011). Cited on page(s) 12

Cherpak, A. (2009) "Carleton University course essay." With permission. Cited on page(s) 23

CIA (2011) *The World Fact Book.* `https://www.cia.gov/library/publications/the-world-factbook/geos/ni.html` (accessed 2011). Cited on page(s) 46

CIOMS and WHO (1993) "International ethical guidelines for biomedical research involving human subjects." *XXVIth CIOMS Conference*, Geneva, CIOMS. Cited on page(s) 46

CIPO (2011) `http://www.cipo.ic.gc.ca/eic/site/cipointernet-internetopic.nsf/eng/Home/` (accessed October 18, 2011). Cited on page(s) 13

CPSO (2010) (College of Physicians and Surgeons of Ontario). "Fetal ultrasound for non-medical reasons." May 2010. `http//www.cpso.on.ca/policies/default.aspx?ID=1606/` (accessed September 2011). Cited on page(s) 28

Coward, H. and G. Hartrick (2000) "Perspectives on health and cultural pluralism: ethics in medical education." *Clin. Invest. Med.*, 261–5. Cited on page(s)

Davidson, A.J. and M. O'Brien (2009) "Ethics and medical research in children." *Pediatric Anesthesia*, August 26, 994–1004. DOI: 10.1111/j.1460-9592.2009.03117.x Cited on page(s) 22, 23

Davis, L. (2002) *Bending Over Backwards: Disability, Dismodernism and Other Difficult Positions.* New York, New York Univ. Press. Cited on page(s) 31

Eisenberg, N., P. Miller and R. Shell (1991) "Prosocial development in adolescence." *Developmental Psychology*, 849–857. DOI: 10.1037/0012-1649.27.5.849 Cited on page(s) 22

Emory, W.C. (1980) *Business Research Methods.* Irwin, 24, 33–34, 20. Cited on page(s) 54, 55

Engineers Canada (2008) "About Engineers Canada." `http://www.engineerscanada.ca/e/en_about.cfm` (accessed August 01, 2011). Cited on page(s) 4

Ennett, C.M., M. Frize and E. Charette (2004) "Improvement and automation of artificial neural networks to estimate medical outcomes." *Medical Eng. and Physics*, 321–328. DOI: 10.1016/j.medengphy.2003.09.005 Cited on page(s) 55

Flaherty, M.P. and D. Struck (2000) "Life by luck of the draw." *The Washington Post*, December 22, A1. Cited on page(s) 45

Frize, M. (1995) "Eradicating harassment in higher education and non-traditional workplaces: a model." *CAASHHE Conference*, Saskatoon, Nov. 1995, 15–18, 43–47. Cited on page(s) 47

Frize, M. (1996) "Teaching ethics & the governance of the profession: Ahead or behind professional practice realities?" *9th Canadian Conference on Engineering Education*, 483–489. Cited on page(s) 36

Frize, M. (2003) "Teaching ethics for bioengineers in the 21st century." *Proceedings of the 25th Annual International Conference of the IEEE EMBS*. Cancun: IEEE EMBS, 3466–3468. Cited on page(s) 36

Frize, M. and N. Yu (2010) "Estimating pre-term birth using a hybrid pattern classification system." *Proc. MEDICON 2010*, Thessaloniki, Greece, May: pp. 893–896. Cited on page(s) 38

Gert, B. (2011) "The definition of morality." April 14. `http://plato.stanford.edu/archives/sum2011/entries/morality-definition/` (accessed August 01, 2011). Cited on page(s) 5

Global Footprint Network (2011) "Global footprint network: Advancing the science of sustainability." April 2011. `http://www.footprintnetwork.org/en/index.php/GFN/page/footprint_basics_overview/` (accessed August 2011). Cited on page(s) 41

Goodman, G. (1989) "The profession of clinical engineering." *Journal of Clinical Engineering*, 27–37. Cited on page(s) 5

Government of Canada (2011) "Panel on Research ethics." April 07. `http://www.pre.ethics.gc.ca/eng/policy-politique/initiatives/tcps2-eptc2/Default/` (accessed August 01, 2011). Cited on page(s) 10

govtrack.us (2006) "H.R. 5641: Safe overseas human testing act." `http://www.govtrack.us/congress/bill.xpd?bill=h109--5641` (accessed September 2011). Cited on page(s) 46

Hand, D. (2007) "Deception and dishonesty with data: fraud in science." *Significance*, March 2007: 22–25. DOI: 10.1111/j.1740-9713.2007.00215.x Cited on page(s) 13

healthpolicycoach.org (2011) "Assisted reproductive technologies." Health Policy Coach. `http://www.healthpolicycoach.org/topics/category/17057-assisted-reproductive-technologies` (accessed September 2011). Cited on page(s) 25

Hinman, L. (2011) "Bioethics, cloning, & reproductive technologies." July 10, 2011. `http://ethics.sandiego.edu/Applied/Bioethics/index.asp` (accessed September 2011). Cited on page(s) 26

Hogle, L.F. (2005) "Enhancement technologies and the body." *Annual Rev. Anthropol.*, 34: 695–716. DOI: 10.1146/annurev.anthro.33.070203.144020 Cited on page(s) 31

Holbrook, J. (2005) "Assessing the science–society relation: The case of the US National Science Foundation's second merit review criterion." *Technology in Society*, 27, 437–451. DOI: 10.1016/j.techsoc.2005.08.001 Cited on page(s) 43, 44

Hvistendahl, M. (2011) "Unnatural selection-choosing boys over girls and the consequences of a world full of men." *Public Affairs*, USA, 250 West 57th Street, Suite 1321, NY, NY 10107. Cited on page(s) 28

Hwang, D.W. (2005) "Patient-specific embryonic stem cells derived from human SCNT blastocysts." *Science*, 308, 1777–83. DOI: 10.1126/science.1112286 Cited on page(s)

Hwang, W.S. et al. (2004) "Evidence of a pluripotent human embryonic stem cell line derived from a cloned blastocyst." *Science*, 303, 1669–74. DOI: 10.1126/science.1094515 Cited on page(s)

IEEE (2011) `http://www.ieee.org/about/corporate/governance/p7--8.html/` (accessed October 18, 2011). Cited on page(s) 10

International Engineering Alliance (2011) "The Washington Accord." `http://www.washingtonaccord.org/Washington-Accord/` (accessed August 01, 2011). Cited on page(s) 4

Joy, B. (2000) "Why the future doesn't need us." *Wired*, April 2000. `http://www.wired.com/wired/archive/8.04/joy.html` (accessed 2011). Cited on page(s) 36

Kluge, Eike-Henner W. (2005) *Readings in Biomedical Ethics: A Canadian Focus*. Scarborough: Prentice-Hall, 2005: 4. Cited on page(s) 5

Kofi Annan (2011) "UN millenium development goals." `http://www.lifebridge.org/UNmillennium.cfm` (accessed August 2011). Cited on page(s)

McNeil, M., I. Varcoe and S. Yearley (1990) *The New Reproductive Technologies*. London: Macmillan Press. Cited on page(s) 26

Miah, A. (2009) *guardian.co.uk*, May 01, 2009. `http://www.guardian.co.uk/science/2009/may/01/body-enhancement-cosmetic-surgery-genetics` (accessed August 01, 2011). Cited on page(s) 32

Mills, K. and C. Fleddermann (2005) "Getting the best from nanotechnology: Approaching social and ethical implications openly and proactively." *IEEE Technology and Society Magazine*, 18–26. DOI: 10.1109/MTAS.2005.1563498 Cited on page(s) 42, 43

Miyaji, N. (1994) "Kokuchi o meguru nihon no ishi no shiseikan, kohen (Japanese doctor's attitudes toward life and death as related to their truth-telling to dying patients, Part 2)." *Taaminarukea*, 497–504. Cited on page(s) 50

Moreno, J.D. (2005) *Is There an Ethicist in the House? On the Cutting Edge of Bioethics.* Bloomington: Indiana University Press. Cited on page(s) 49

Morioka, M. (1995) "Bioethics and Japanese culture: Brain death, patients' rights, and cultural factors." *Eubios Journal of Asian and International Bioethics*, 5, 87–90. Cited on page(s) 49, 50

MSNBC (2005) "Charges of fake research hit new high: Doctors accused of making up data in medical studies." July 10, 2005. `http://www.msnbc.msn.com/id/8474936/` (accessed June 2011). Cited on page(s)

National Society of Professional Engineers (2011) "Licensure." `http://www.nspe.org/Licensure/HowtoGetLicensed/index.html` (accessed August 01, 2011). Cited on page(s) 4

NIH (2011) `http://grants.nih.gov/grants/policy/air/index.htm/` (accessed October 18, 2011). Cited on page(s) 10, 24

Office of Research Integrity (2006) "Definition of research misconduct." December 06, 2006. `http://ori.hhs.gov/misconduct/definition_misconduct.shtml` (accessed August 01, 2011). Cited on page(s) 12

Office of Research Integrety (2005) "Press Release - Dr. Eric T. Poehlman." March 17, 2005. `http://ori.hhs.gov/misconduct/cases/press_release_poehlman.shtml` (accessed August 2011). Cited on page(s)

Ondrusek, N, P. Pencharz and G. Koren (1998) "Empirical examination of the ability of children to consent to clinical research." *J. Med. Ethics*, 158–165. DOI: 10.1136/jme.24.3.158 Cited on page(s) 22

PEO (Professional Engineers Ontario) (2011) `www.peo.on.ca/` (accessed October 18, 2011). Cited on page(s) 47

Pinxten, W., K. Dierickx and H. Nys (2009) "Ethical principles and legal requirements for pediatric research in the EU: An analysis of the European normative and legal framework surrounding pediatric clinical trials." *European Journal of Pediatrics*, 1225–1234. DOI: 10.1007/s00431-008-0915-7 Cited on page(s) 22, 23

Planet Powai (2011) "Saki Naka doctor caught red handed in sex determination case." July 11, 2011. `http://www.planetpowai.com/news/1707201104.htm` (accessed August 2011). Cited on page(s) 28

Presidential Council for the Study of Bioethics (2003) "Beyond therapy: Biotechnology and the pursuit of happiness." `http://www.bioethics.gov/reports/beyondtherapy` Cited on page(s) 31

Psychology Wiki (2011) "Theories." `http://psychology.wikia.com/wiki/Theories` (accessed August 2011). DOI: 10.2307/2095660 Cited on page(s) 56

Quinn, M. (2005) *Ethics for the Information Age.* Boston: Pearson/Addison-Wesley, 53–55, 52–62, 64, 62–67, 67–72, 72–76, 76–84 , 81–84, 78–74. Cited on page(s) 6, 7, 8, 9

Ready, T. (2001) "Pfizer in "unethical" trial suit." *Nature Medicine*, v. 7, 1077. DOI: 10.1038/89847 Cited on page(s) 46

Ready, T. (2000) "Placebo trials deemed unethical." *Nature Medicine*, v. 6, 1198. DOI: 10.1038/72191 Cited on page(s) 46

Reproductive Health Technologies Project (2011) "Prenatal testing." `http://www.rhtp.org/fertility/prenatal/default.asp` (accessed July 2011). DOI: 10.1016/0028-2243(80)90085-4 Cited on page(s) 27

Resnick and Shamoo (2003) *Responsible Conduct of Research.* New York: Oxford University Press, 261–2, 80. Cited on page(s) 3, 14, 24, 29

Robison, W.L. (2010) "Bioinformatics and privacy." *Ethics in Biology, Engineering & Medicine - An International Journal*, vol. 1(1), 9–17. DOI: 10.1615/EthicsBiologyEngMed.v1.i1.30 Cited on page(s) 13

Rosser, S. (1989) "Re-visioning clinical research: Gender and the ethics of experimental design." *Feminist Ethics & Medicine*, 125–139. Cited on page(s) 48

Royal Commission on New Reproductive Technologies (1993) "Social values and attitudes surrounding new reproductive technologies." Canada. `http://publications.gc.ca/site/eng/39774/publication.html` Cited on page(s) 27

Saha, S. and P.S. Saha (1986) "Ethical responsibilities of the clinical engineer." *Journal of Clinical Engineering*, 17–25. Cited on page(s) 5

Sarewitz, D. (1996) *Frontiers of Illusion: Science, Technology, and the Politics of Progress.* Philadelphia: Temple University Press, 19. Cited on page(s) 43

Shrader-Frechette, K.S. (1994) *Ethics of Scientific Research.* Lanham: Rowman & Littlefield Publishers, Inc., 26–36. Cited on page(s) 15, 24

Specht, J. (2000) "Centre for the study of technology and society." `http://www.tecsoc.org/innovate/focusbilljoy.htm` (accessed August 2011). Cited on page(s) 35

Stephens, J. (2006) "Panel faults Pfizer in '96 clinical trial in Nigeria: Unapproved drug tested on children." *Washington Post*, May 07, 2006. `http://www.washingtonpost.com/wp-dyn/content/article/2006/05/06/AR2006050601338.html` (accessed August 2011). Cited on page(s) 46

Stephens, J. (2007) "Pfizer faces new charges over Nigerian drug test." *Washington Post*, June 02, 2007. `http://www.washingtonpost.com/wp-dyn/content/article/2007/06/01/AR2007060102197.html` (accessed August 2011). Cited on page(s)

Taylor, D. (2011) `http://www.stemcell.umn.edu/faculty/Taylor_D/home.html` (accessed October 2011). Cited on page(s) 34

The UK Technology Education Centre (2011) "Technological determinism." `http://atschool.eduweb.co.uk/trinity/determin.html` (accessed August 2011). Cited on page(s) 35

Tong, R. (2007) *New Perspectives in Healthcare Ethics*. Upper Saddle River: Pearson Education Inc., 29, 30, 31–33, 26, 25. Cited on page(s) 15, 16, 23

UN (2006) "Millennium Project." `http://www.unmillenniumproject.org/goals/index.htm` (accessed August 2001). Cited on page(s) 42, 44

United Nations (2011) "The millennium development goals report." `http://www.un.org/millenniumgoals/pdf/%282011_E%29%20MDG%20Report%202011_Book%20LR.pdf` (accessed August 2011). Cited on page(s) 45

UNDP (1989) United Nations Development Program. Cited on page(s) 41

University of California (2008) "New method devised for protecting private data." `http://www.universityofcalifornia.edu/news/article/17676` (accessed September 2011). Cited on page(s) 40

University of Ottawa (2011) "Consent process and forms." `http://www.research.uottawa.ca/ethics/consent.html` (accessed August 01, 2011). Cited on page(s) 19

USPTO (2011) United States Patent and Trademark Office. Cited on page(s)

Voigt, H.F. and D.M. Ehrmann (2010) "The Ethical Code for Medical and Biological Engineers Should Preclude theur Role in Judicial Executions." Ethics in Biol., Eng., and Med. 1(1):43-52. Cited on page(s) 11

Voigt, H.F. (2010) "Editorial: A Need for a Universal Code of Ethics for Medical and Biological Engineers." Ethics in Biol., Eng., and Med. 1(2):79-81. Cited on page(s) 11

Walker, S., F. Mussen and S. Salek (2009) *Benefit-Risk Appraisal of Medicine: A systematic Approach to Decision-Making*. Chippenham: Wiley-Blackwell. Cited on page(s) 21

Weiss, C. (2005) "Science, technology and international relation." *Technology in Society*, 27, 295–313. DOI: 10.1016/j.techsoc.2005.04.004 Cited on page(s) 36, 37

Weyand, S., M. Frize, E. Bariciak and S. Dunn (2011) "Development and usability testing of a combined physician-parent decision-support tool for the neonatal intensive care unit." *Proc. IEEE/EMBC 2011*, Boston. Cited on page(s) 21, 39

Wikipedia (2011a) "Bias." `http://en.wikipedia.org/wiki/bias/` (accessed October 18, 2011). Cited on page(s) 55

Wikipedia (2011b) "Conflict of Interest." `http://en.wikipedia.org/wiki/Conflict_of_interest` (accessed August 2001). Cited on page(s) 14

Wikipedia (2011c) "Corruption." `http://en.wikipedia.org/wiki/Corruption` (accessed August 28, 2011). Cited on page(s) 12

Wikipedia (2011d) "Fraud." `http://en.wikipedia.org/wiki/Fraud` (accessed August 01, 2011). Cited on page(s) 12

Wikipedia (2011e) "Hypothesis." `http://en.wikipedia.org/wiki/Hypothesis` (accessed August 2011). Cited on page(s) 55

Wikipedia (2011f) "Intellectual Property." `http://en.wikipedia.org/wiki/intellectual_property` (accessed October 18, 2011). Cited on page(s) 13

Wikipedia (2011g) "Legal status of Internet pornography." `http://en.wikipedia.org/wiki/Legal_status_of_Internet_pornography` (accessed August 2011). Cited on page(s) 40

Wikipedia (2011h) "Orlan." `http://en.wikipedia.org/wiki/Orlan` (accessed August 01, 2011). Cited on page(s) 32

Wikipedia (2011i) "Ownership." `http://en.wikipedia.org/wiki/Ownership#Intellectual_property` (accessed August 2011). Cited on page(s) 13

Wikipedia (2011j) "Oscar Pistorius." `http://en.wikipedia.org/wiki/oscar_pistorius` (accessed October 18, 2011). Cited on page(s) 33

Wikipedia (2011k) "Physical law." `http://en.wikipedia.org/wiki/Physical_law` (accessed August 2011). Cited on page(s) 56

Wikipedia (2011l) "Reductionism." `http://en.wikipedia.org/wiki/Reductionism` (accessed August 2011). Cited on page(s) 56

Wikipedia (2011m) "Society." `http://en.wikipedia.org/wiki/Society` (accessed August 2011). Cited on page(s) 32, 36

Wikipedia (2011n) "Technological determinism." September 2011. `http://en.wikipedia.org/wiki/Technological_determinism`. Cited on page(s) 35

Wise, J. (2001) "Pfizer accused of testing new drug without ethical approval." *British Medical Journal*, v. 322, 194. DOI: 10.1136/bmj.322.7280.194 Cited on page(s) 46

WMA (2004) "Regulations and ethical guidelines." `http://ohsr.od.nih.gov/guidelines/helsinki.html` (accessed August 2011). Cited on page(s) 11, 46

Wolfe, L. (1997) "Unethical trials of interventions to reduce perinatal transmission of the human immunodeficiency virus in developing countries." *N. Engl. J. Med.*, 853–6. Cited on page(s) 45

World Medical Association (2004) "World Medical Association Declaration of Helsinki." `http://www.wma.net/e/policy/b3.html` (accessed March 01, 2007). Cited on page(s) 11

Author's Biography

MONIQUE FRIZE

Monique Frize is a Distinguished Professor at Carleton and Professor Emerita at University of Ottawa. She was a biomedical engineer for 18 years in hospitals (1971-1989) and a professor in electrical and biomedical engineering since 1989. Monique Frize published over 200 papers in journals and conference proceedings on artificial intelligence in medicine, infrared imaging, ethics and women in engineering and science. She is Senior Member of IEEE, Fellow of the Canadian Academy of Engineering (1992), Fellow of Engineers Canada (2010), Officer of the Order of Canada (1993) and recipient of the 2010 Gold Medal from Professional Engineers Ontario and the Ontario Society of Professional Engineers. On January 1, 2012, she will be a Fellow of IEEE for her work in clinical engineering and engineering education. She received five honorary doctorates in Canadian universities since 1992. Her book: *The Bold and the Brave: A history of women in science and engineering* was released by University of Ottawa Press in November 2009.